Statistical Issues in Allocating Funds by Formula

Panel on Formula Allocations

Thomas A. Louis, Thomas B. Jabine, and Marisa A. Gerstein, *Editors*

Committee on National Statistics

Division of Behavioral and Social Sciences and Education

NATIONAL RESEARCH COUNCIL
OF THE NATIONAL ACADEMIES

THE NATIONAL ACADEMIES PRESS
Washington, D.C.
www.nap.edu

THE NATIONAL ACADEMIES PRESS   500 Fifth Street, N.W.   Washington, DC 20001

NOTICE: The project that is the subject of this report was approved by the Governing Board of the National Research Council, whose members are drawn from the councils of the National Academy of Sciences, the National Academy of Engineering, and the Institute of Medicine. The members of the committee responsible for the report were chosen for their special competences and with regard for appropriate balance.

This study was supported by Contract/Grant No. RN 96131001 between the National Academy of Sciences and the U.S. Department of Education. Support of the work of the Committee on National Statistics is supported by a consortium of federal agencies through a grant from the National Science Foundation (Number SBR-9709489). Any opinions, findings, conclusions, or recommendations expressed in this publication are those of the author(s) and do not necessarily reflect the views of the organizations or agencies that provided support for the project.

International Standard Book Number 0-309-08710-4
Library of Congress Catalog Card Number 2002116513

Additional copies of this report are available from the National Academies Press, 500 Fifth Street, N.W., Lockbox 285, Washington, DC 20055; (800) 624-6242 or (202) 334-3313 (in the Washington metropolitan area); Internet, http://www.nap.edu

Suggested citation: National Research Council (2003). *Statistical Issues in Allocating Funds by Formula*. Panel on Formula Allocations. Thomas A. Louis, Thomas B. Jabine, and Marisa A. Gerstein, Editors. Committee on National Statistics, Division of Behavioral and Social Sciences and Education. Washington, DC: The National Academies Press.

THE NATIONAL ACADEMIES
Advisers to the Nation on Science, Engineering, and Medicine

The **National Academy of Sciences** is a private, nonprofit, self-perpetuating society of distinguished scholars engaged in scientific and engineering research, dedicated to the furtherance of science and technology and to their use for the general welfare. Upon the authority of the charter granted to it by the Congress in 1863, the Academy has a mandate that requires it to advise the federal government on scientific and technical matters. Dr. Bruce M. Alberts is president of the National Academy of Sciences.

The **National Academy of Engineering** was established in 1964, under the charter of the National Academy of Sciences, as a parallel organization of outstanding engineers. It is autonomous in its administration and in the selection of its members, sharing with the National Academy of Sciences the responsibility for advising the federal government. The National Academy of Engineering also sponsors engineering programs aimed at meeting national needs, encourages education and research, and recognizes the superior achievements of engineers. Dr. Wm. A. Wulf is president of the National Academy of Engineering.

The **Institute of Medicine** was established in 1970 by the National Academy of Sciences to secure the services of eminent members of appropriate professions in the examination of policy matters pertaining to the health of the public. The Institute acts under the responsibility given to the National Academy of Sciences by its congressional charter to be an adviser to the federal government and, upon its own initiative, to identify issues of medical care, research, and education. Dr. Harvey V. Fineberg is president of the Institute of Medicine.

The **National Research Council** was organized by the National Academy of Sciences in 1916 to associate the broad community of science and technology with the Academy's purposes of furthering knowledge and advising the federal government. Functioning in accordance with general policies determined by the Academy, the Council has become the principal operating agency of both the National Academy of Sciences and the National Academy of Engineering in providing services to the government, the public, and the scientific and engineering communities. The Council is administered jointly by both Academies and the Institute of Medicine. Dr. Bruce M. Alberts and Dr. Wm. A. Wulf are chair and vice chair, respectively, of the National Research Council.

www.national-academies.org

Acknowledgments

Many people have contributed their time, expertise, and resources to the panel. The National Center for Education Statistics of the U.S. Department of Education provided financial and program support, and we thank Daniel Kasprzyk for his interest and participation on their behalf. Andrew White, director of the Committee on National Statistics (CNSTAT), developed institutional and financial support for the panel.

Many individuals provided important comments and suggestions at panel meetings (see Appendix E). We thank them for their time and insights. We thank the authors of the papers prepared for the panel, which appear in a special issue of the *Journal of Official Statistics*. Dawn Aldridge, U.S. Department of Agriculture; Paul Sanders Brown, U.S. Department of Education; Jean-Francois Carbonneau, Statistics Canada; John L. Czajka, Mathematica Policy Research, Inc.; Thomas Downes, Tufts University; Thomas B. Jabine, statistical consultant; James A. Kadamus, New York State Education Department; Sean Keenan, Finance Canada; Dan Melnick, Dan Melnick Research; Thomas Pogue, University of Iowa; Allen L. Schirm, Mathematica Policy Research, Inc.; Felizardo B. Suzara, United Nations Statistics Division; Michelle Taylor, Finance Canada; and Alan M. Zaslavsky, Harvard University. We also thank the reviewers of those papers: David Betson, University of Notre Dame; Linda Bilheimer, Robert Wood Johnson Foundation; Hamilton Lankford, Stanford University; David McMillen, House Government Reform Committee; Wayne Riddle, Congressional Research Service; William Seltzer, Fordham University;

Robert Tannenwald, Federal Reserve Bank of Boston; and Jim Wyckoff, State University of New York, Albany.

We have greatly benefited from the contributions of the late Wray Smith, who chaired the federal interagency committee that prepared the 1978 *Report on Statistics for Allocation of Funds.* He also wrote and presented the keynote paper for the Workshop on Formulas for Allocating Program Funds, held by CNSTAT in April 2000.

I want to personally thank panel members, who donated their time and expertise, and give special thanks to Thomas Jabine, consultant to the panel, whose thoughtful guidance, broad and deep knowledge, and good old-fashioned hard work were instrumental in making this report possible. Marisa Gerstein, research assistant, contributed in myriad fashions, organizing, researching, writing, editing, and generally keeping us upbeat. Virginia de Wolf, study director, and Maria Alejandro, project assistant, ensured success for meetings and provided other key organizational contributions. Christopher Mackie, CNSTAT program officer, can and did play all positions and frequently pitched in.

This report has been reviewed in draft form by individuals chosen for their diverse perspectives and technical expertise, in accordance with procedures approved by the Report Review Committee of the National Research Council. The purpose of this independent review is to provide candid and critical comments that will assist the institution in making the published report as sound as possible and to ensure that the report meets institutional standards for objectivity, evidence, and responsiveness to the study charge. The review comments and draft manuscript remain confidential to protect the integrity of the deliberative process.

We thank the following individuals for their participation in the review of this report: Henry J. Aaron, The Brookings Institution; Jerry C. Fastrup, Government Accounting Office; Thomas Gabe, Congressional Research Service; Cynthia Long, Food and Nutrition Services, U.S. Department of Agriculture; Fritz Scheuren, Urban Institute; James Wyckoff, Rockefeller College, University at Albany, SUNY; and Alan M. Zaslavsky, Harvard Medical School.

Although the reviewers listed above have provided many constructive comments and suggestions, they were not asked to endorse the conclusions or recommendations nor did they see the final draft of the report before its release. The review of this report was overseen by John C. Bailar, Professor Emeritus, University of Chicago, and Joel Greenhouse, Department of Statistics, Carnegie Mellon University. Appointed by the National Research

Council, they were responsible for making certain that an independent examination of this report was carried out in accordance with institutional procedures and that all review comments were carefully considered. Responsibility for the final content of this report rests entirely with the authoring panel and the institution.

Thomas A. Louis, *Chair*
Panel on Formula Allocations

Contents

Preface

Each year in the United States statistical formulas are used to allocate large amounts of federal funds to state and local governments in programs designed to meet a wide spectrum of economic and social objectives, such as improving educational outcomes and accessibility to medical care. Many of the programs are designed to equalize the fiscal capacities of the recipient units of government to address identified needs. Furthermore, many states use formulas to distribute aid to local governments.

There is a long history of attention to issues associated with allocating funds by formula, though surprisingly little of it has come from statisticians. To reenergize discussion on these matters, the Committee on National Statistics (CNSTAT) convened, in April 2000, a two-day Workshop on Formulas for Allocating Program Funds. Drawing examples from four major U.S. programs, the workshop focused on statistical issues that arise in the development and use of formulas for allocating federal funds to state and local governments. Presenters and other workshop participants included formula allocation program managers, economists, statisticians, and demographers from federal and state government agencies, universities, and independent research organizations.

The workshop was a direct outgrowth of a previous study by the Panel on Estimates of Poverty for Small Geographic Areas. That panel, established under a 1994 act of Congress, began its work with a specific mission: to evaluate the suitability of the U.S. Census Bureau's small-area estimates of the number of poor school-age children for use in the allocation of funds to counties and school districts under Title I of the Elementary and Secondary Education Act. In considering that panel's conclusions, CNSTAT decided that it was important to explore these issues in a broader context, starting with the workshop and proceeding to a more comprehensive panel study.

The Panel on Formula Allocations was formed in January 2001 with sponsorship by the National Center for Educational Statistics and also, in part, by the many U.S. federal statistical agencies that support CNSTAT through the National Science Foundation. The panel's charge was to refine and follow up on the important issues identified in the workshop, conduct case studies and methodological investigations, obtain input from individuals who design and implement programs using formula allocation, and to develop findings, conclusions, and recommendations relating to these issues. Initially, the panel concentrated its efforts on technical and conceptual statistical issues but, as the study progressed, found it essential to embed the statistical issues in a broader political and policy framework.

In July 2001 the panel issued *Choosing the Right Formula: Initial Report* (National Research Council, 2001), which featured the report of the April 2000 workshop, highlighted key issues identified by the panel, and communicated its work plan. In conjunction with its own work, the panel also commissioned a series of papers—some devoted to case studies and others to an examination of the goals of formula allocation programs, the specific roles played by formulas, and the statistical features of allocation formulas and processes. Several of these papers were presented at the panel's December 2001 meeting, and all of them are assembled as a special issue of the *Journal of Official Statistics* (September 2002). The journal issue is intended to serve as a companion to this, the panel's final report, and is a key component of the panel's work toward fulfilling its charge.

I am confident that you will find the panel's final report interesting and informative. Using allocation formulas to advance legislative aims has become a widely employed policy tool. The panel hopes that this report will be helpful in informing the process in the future when allocation formulae are being crafted in legislation.

John E. Rolph, *Chair*
Committee on National Statistics

Executive Summary

Mathematical formulas are used to allocate more than $250 billion of federal funds annually to state and local governments via more than 180 grant-in-aid programs. These programs promote a wide spectrum of economic and social objectives, such as improving educational outcomes and increasing accessibility to medical care, and many are designed to compensate for differences in fiscal capacity that affect governments' abilities to address identified needs. Large amounts of state revenues are also distributed through formula allocation programs to counties, cities, and other jurisdictions.

The essential feature of a formula allocation program is that fund distribution is determined by the application of a formula that uses statistical information to calculate or estimate the values of its inputs. The allocation process consists of a basic calculation using a mathematical formula or algorithm; it often includes adjustments that place constraints on levels or shares (percentages of the total allocation) or on changes in levels or shares. Many programs use official statistics as inputs in the estimation of the central formula components—need, capacity, and effort. The kinds of data used vary widely: total population, population by age group, per capita income, and proportion of persons with family income below the poverty line are a few examples. In several instances, data collection programs were initiated or expanded specifically to provide data needed for funds allocation.

Allocation formulas are designed with one or more objectives and are developed in the context of a complex political process. In addition to providing a mechanism for addressing changes in need and other formula components without Congress having to revisit the issue annually, formula-based allocations can help build consensus for and the credibility of a program. Use of a formula (rather than a possibly arbitrary specification of amounts to be given to recipient jurisdictions) facilitates informed debate and a degree of transparency about the allocation process by providing documentation of assumptions and computations. Furthermore, a formula offers legislators an effective way of explaining the allocation process to their constituents. However, when funds are allocated according to a formula, there is no guarantee that objectives will be fully met. In particular, properties of data sources and statistical procedures used to produce formula inputs can interact in complex ways with formula features to produce consequences that may not have been anticipated or intended.

This report identifies key issues concerning the design and use of formulas for fund allocation and advances recommendations for improving the process. Most of the panel's conclusions and recommendations fall into one of two overlapping sets: the first pertains to issues created by the interaction between the political process and formula design, the second to internal design and data issues more narrowly. In addition, the panel makes two specific programmatic recommendations.

In the first area, the panel emphasizes the importance of finding the proper balance between legislative control and program agency autonomy. At one extreme, the basic formula, the variables used to estimate its components, the data sources, and the special features would be fully specified in legislation. At the other extreme, the legislation would define the general objectives of the program, and the program agency would develop the specific formula and allocation procedures. The panel calls on formula allocation program designers in both the legislative and executive branches to be aware of and to evaluate the potential for behavioral responses by the funded jurisdictions aimed at influencing input data or other factors that affect calculated funding levels.

In addressing the design and data issues, the panel begins with guidance about how to evaluate trade-offs in timeliness, quality, costs, and other factors that are inherent in the selection of the variables used as formula components and in the data sources and methods used to estimate them. The panel strongly recommends that careful evaluation of the potential effects on allocations associated with special formula features—such as hold-harmless provisions, thresholds, and minimums—be part of any formula review protocol. Evaluations should focus on how use of these features may cause unintended departure of allocations from estimated need levels. In this context, use of simulations during program development and reauthorization review should be expanded, and they should be used to examine both cross-sectional and longitudinal allocation patterns. Simulations are also recommended as a means for exploring cross-cutting issues, such as those that arise when assessing alternative measures of fiscal capacity, and the relative merits of using hold-harmless provisions or moving averages to dampen the effects of large changes in the formula inputs.

The panel makes targeted recommendations to the General Services Administration (GSA) and the Office of Management and Budget (OMB). The GSA's *Catalog of Federal Domestic Assistance* is an excellent tool, but several specific changes could improve its value for users seeking information about federal allocation programs. The panel also recommends that OMB establish a standing Interagency Committee on Formula Allocations charged with developing improved simulation and quality-control techniques for use in formula design and fund allocation procedures. Appendix E of this report contains a template that could be used by the interagency committee to develop a handbook that would serve as an introduction to underlying concepts and practical considerations in the use of formula-based fund allocation. It would be valuable to those in the legislative and executive branches who are involved in the design and operation of formula allocation programs and could be used in training programs for various audiences.

Chapter 9 presents the panel's recommendations. They are supported by the discussion of statistical and political issues throughout the report and also by a set of papers, commissioned by the panel, which appear in the *Journal of Official Statistics* (September 2002).

1

Introduction

THE USE OF FORMULAS TO ALLOCATE FUNDS

Each year, formulas are used to allocate large amounts of federal funds to state and local governments via federal programs designed to meet a wide spectrum of economic and social objectives. These programs address societal goals, such as improving educational outcomes and accessibility of medical care, and some of them are designed to equalize fiscal capacity to address identified needs.

Funds distributed by federal formula allocation grants have more than doubled in real terms over the past 25 years. A 1975 study estimated that $35.6 billion was allocated under grants using population or per capita income as formula components. By fiscal year (FY) 2000, Medicaid was by far the largest formula allocation program, with $111.1 billion disbursed. Federal-Aid Highway Program grants were the next largest, with $25.9 billion, followed by allocations under the Temporary Assistance for Needy Families program (TANF), with $19.1 billion. The December 2001 update of the Catalog of Federal Domestic Assistance showed 180 federal formula allocation programs with FY 2000 obligations totaling approximately $262.3 billion. Table 1-1 lists the 11 largest programs for fiscal years 1999 and 2000 (see Appendix B for more details).

Formula allocation programs are characterized by the allocation of money to states or their subdivisions in accordance with a distribution formula prescribed by law or administrative regulation, for activities of a

TABLE 1-1 The 11 Largest Federal Programs, Fiscal Years 1999 and 2000

Program	Obligations ($billions)		Rank	
	FY 1999	FY 2000	FY 1999	FY 2000
Medical Assistance Program (Medicaid)	111.1	121.8	1	1
Highway Planning and Construction	26.2	25.9	2	2
Temporary Assistance for Needy Families	18.8	19.1	3	3
Title I Education	7.7	7.9	4	4
National School Lunch Program (food grant portion)	5.5	5.6	5	5
Special Education Grants to States	4.3	5.0	6	6
State Children's Health Insurance	4.2	4.3	7	8
Foster Care Title IV-E	4.0	4.5	8	7
Community Development Block Grants	3.0	3.0	9	10
WIC (food grant portion)	2.9	2.8	10	11
Public Housing Capital Fund	2.2	3.9	14	9
Total for 11 largest programs	187.7	201.0		
Total for all formula allocation programs	245.9	262.3		

Source: Catalog of Federal Domestic Assistance <http://www.cfda.gov>

continuing nature not confined to a specific project. For some programs, the distribution formula used is a closed mathematical expression; for others, iterative processes are used to arrive at the final allocations. Block grant programs are a subset of formula allocation programs in which the recipient jurisdiction has broad discretion for the application of funds received in support of such programs as community development or the prevention and treatment of substance abuse, which are specified in the authorizing legislation. Matching grant programs, such as Medicaid and certain transportation programs, require that the recipient state provide a matching percentage of funds from state sources.

Allocation formulas are usually designed with one or more objectives: to distribute funds to recipient governments in proportion to some measure of program need, to equalize their fiscal capacities to meet program needs, or to influence their spending decisions. They are developed in the context of a complex political process. Use of a formula (rather than a specification of the amount to be given to each recipient jurisdiction) facilitates informed debate and a degree of transparency about the allocation process by providing documentation of assumptions and computations. Furthermore, a formula offers legislators an effective way of explaining the allocation process to their constituents. However, when funds are allocated according to a formula, there is no guarantee that objectives will be fully met. In particular, properties of data sources and statistical procedures used to produce formula inputs can interact in complex ways with formula features to produce consequences that may not have been anticipated or intended. Sometimes these problems arise when a formula developed in authorizing legislation is modified by provisions introduced in the appropriations process.

The use of formulas to allocate federal funds to subordinate jurisdictions is part of a broader process of government-to-government transfer of funds. Uses of such funds by the recipients may be unrestricted (as in the General Revenue Sharing Program of the 1970s and 1980s and the Canadian Equalization Program), or they may be limited to specific purposes, for example, to provide medical care and improve the education of children in poor families, assist persons to end their dependence on welfare, revitalize economically depressed cities and neighborhoods, or provide assistance to localities disproportionately affected by HIV infection. At the federal level, the U.S. Congress determines how much money will be distributed and for what purposes. For some programs, Congress appropriates a fixed total amount each year to be allocated among states or other recipients; for others, such as Medicaid, amounts may be specified as a certain proportion of all qualified expenditures by a state or other jurisdiction. In the former case, a formula dictates how much of the total goes to each recipient; in the latter case, a formula determines what proportion of each jurisdiction's amount will be matched by the federal government.

In a few programs, federal funds are allocated directly to jurisdictions below the state level, such as school districts. In most federal formula allocation programs, however, state agencies take responsibility for within-state distribution of the funds allocated to them, subject to program regulations and various kinds of audits and reviews. Importantly, states use formulas to distribute a substantial amount of their own funds. For example, state funds provided for public education substantially exceed those provided by the federal government.

Governments and international organizations use mathematical and statistical formulas for many purposes besides the allocation of funds to subordinate jurisdictions. Examples include the apportionment of seats in the U.S. House of Representatives, judicial determination of amounts awarded for child support (National Research Council, 2001:34-35), payments to health plans in the Federal Employee Health Benefits Program, and "taxes" on federal agency appropriations to cover overhead expenses. However, to make its task more manageable and to intensify its focus, the panel decided to restrict attention to formulas used to allocate funds to government units, with primary, but not unique, focus on allocations by the U.S. federal government.

AN OVERVIEW OF FUND ALLOCATION PROGRAMS

The use of formulas to allocate federal and state funds to subordinate jurisdictions can be traced back to the 19th century. The Morrill Act of 1862 allotted to each state 30,000 acres of public land for each of its senators and representatives in Congress. The land was to be sold and the proceeds used to establish the institutions of higher learning that came to be known as the land grant colleges. Although originally started as agricultural and technical schools, many land grant colleges grew, with additional state aid, into large public universities, which over the years have educated millions of Americans.

Extensive use of formula grants to states did not begin until the 20th century. The Federal-Aid Road Act of 1916, the forerunner of today's Federal-Aid Highway Program, established important precedents for provision of federal assistance through state and local governments, requirements for recipient governments to expend matching funds, and the use of formulas to determine the distribution of program funds (Weingroff, 2001). The broad category of intergovernmental expenditures to state and local governments grew from $2 million in 1902, to $118 million in 1922, to $2.371 billion in 1950 (U.S. Bureau of the Census, 1975).

Growth continued in the second half of the century. A broad-based grant program to provide school aid to federally impacted areas was enacted in 1950, and the first block grant program, comprehensive health grants, was established in 1962 (Break, 1980). In 1965, Title I of the Elementary and Secondary Education Act established a formula-based allocation of federal funds designed to improve educational opportunities for school-age children from poor families.

The General Revenue Sharing Program, enacted in 1972, represented an attempt to equalize fiscal capacity across jurisdictions by providing federal funds for unrestricted use by the recipients. Similar programs of unrestricted grants, which continue to the present, already existed in Canada and Australia. Under general revenue sharing, federal funds were allocated to approximately 39,000 local jurisdictions using a formula based on tax effort, population, and per capita income. The program continued through 1986, but during a period of budgetary stringency was not reauthorized by Congress. However, categorical and block grant programs continued to grow. Today, formulas are used to allocate well over $250 billion of federal funds annually to state and local governments via approximately 180 federal programs designed to meet a wide spectrum of economic and social objectives.[1] Large amounts of state revenues are distributed to cities, counties, and other local governments by way of formula allocation programs.

The essence of a formula allocation program is that the amounts of money to be allocated through the program are calculated by a specified formula that requires various statistical inputs and other information. The formula may be expressed as a mathematical equation that directly provides the amounts, given the appropriate input quantities, or as an algorithm for calculating the amounts. Frequently the formula consists of a basic calculation (an equation or an algorithm) that provides initial amounts, followed by a series of additional rules or adjustments that amend these amounts through constraints on their level or change over time. In some cases, the formula produces the actual dollar amounts to be allocated; in other cases, it produces the shares of an available total amount to go to each jurisdiction; in yet other cases (e.g., Medicaid), it provides a matching percentage for supplementing expenditures being made by the recipient jurisdiction from its own funds.

Formula allocation programs also vary in terms of the uses to which funds can be put. Many are directed at alleviating particular socioeconomic problems, with the use of funds restricted to that purpose. Of course, a recipient jurisdiction might, on receiving directed federal funding, be tempted to reduce its own expenditure in the problem area, thus freeing up funds for other purposes.[2] Some programs, such as Medicaid, try to

---

[1]Appendix C describes how the panel used data from the Catalog of Federal Domestic Assistance to identify the program universe studied and provides more precise dollar figures for allocations.

[2]There is an extensive literature on this subject, which the panel has chosen not to summarize in this report. For additional information on fiscal substitution, see generally: Advisory Commission on Intergovernmental Relations (1977); Gramlich and Galper (1973); Lindsey and Steinberg (1990); U.S. General Accounting Office (1996).

discourage such substitution by making allocations proportional to the recipients' expenditures for the program. Others, such as TANF, require the recipient to spend at least a specified amount—for example, some proportion of its federal grant—from its own funds for the program.

Why Formulas?

The potential benefits of a formula-driven approach can be understood by considering the alternatives. In the absence of a formula, one of two scenarios may arise. If responsibility for the distribution of funds remains with Congress, members would have to negotiate an allocation afresh for each fiscal period. If the allocation of funding has been delegated to program administrators, they would have to deal with the pressures from recipient jurisdictions in making their allocation decisions each year. In either case it makes good sense that, if an initial round of allocations has been negotiated on the basis of certain principles or rationale, those principles or rationale be embedded in a formula so that equivalent results may be reproduced (with necessary changes being made) in subsequent years. Of course, the formula-based approach is not perfect, and much of this report is concerned with identifying and dealing with problems that arise in the design and implementation of formula-based programs.

The crucial concept in the previous paragraph—that allocations are based on some principles or rationale—is critical to one's view of formula allocation programs. Two extreme views of formula allocation programs can be contrasted. At one extreme is the view that underlying any formula allocation program there exists a clear set of principles that define how funds should be allocated. For example, the funds a jurisdiction receives should rise with the size of its program caseload, fall with its capacity to raise revenue itself, and perhaps rise with the relative cost of providing the program in that jurisdiction. Even with such principles defined, there is still plenty of room to negotiate formulas in terms of, for example, how exactly caseload, capacity, and cost are to be measured, what weights they are to be given, and how steeply funds should rise or fall with each of these measures. But the existence of such principles or objectives at least provides criteria for assessing how well a given formula program performs, as well as for judging the value of proposed improvements to a formula.

At the other extreme is the view that program allocations are purely the result of political negotiations, that political trade-offs may be made across unrelated programs and issues, and that, while principles may have underlined initial negotiating positions, these will not be recognizable in final allocations. Under this scenario, the use of a formula that yields the negotiated results becomes little more than cover to provide pseudorationality for the results of such a process. No basis for judging improvements to the formula exists under this scenario.

Each of these views is extreme, and reality undoubtedly occupies a middle ground. But, when designing, assessing, or revising formula allocation programs, it is important to recognize the conflicting pressures produced by these alternative views of the role of formulas in the allocation of funds.

Alternative Approaches to Fund Allocation

Paralleling this range of views on how formula allocations come about are various approaches to it. In some cases, the dollar amounts (or shares) to be provided to each jurisdiction are specified in the legislation itself with no pretense of a formula. For example, several years ago in the State Capitalization Grants Program of the Environmental Protection Agency, a formula based on a periodic survey of clean water needs was replaced by legislated shares. In the majority of formula allocation programs, the formula is specified in legislation and leaves little or no discretion to the program administrators to influence the method of calculation. For some of these programs, for example, the State Children's Health Insurance Program (SCHIP), the data sources for each formula element are specified in precise detail in the legislation, leaving the program agency with even less discretion to decide how the formula inputs should be derived. For others, such as Title I education, the program agency has been given substantial leeway in developing estimates of formula components. In a few cases, only the program goals are included in legislation and the formula is developed by the administering program agency. For example, for the food grant portion of the Special Supplemental Nutrition Program for Women, Infants, and Children (the WIC program), Congress established the general objectives in legislation and left the program agency, the U.S. Department of Agriculture's Food and Nutrition Service, to develop the allocation formula, which is set out in regulations published by the agency.

Development and Administration of Formula Programs

Several key players have important roles in the development and administration of formula allocation programs. Congress, as the legislative authority, plays a determining role. Funding through formula allocations represents just one of several methods of funding programs. The funding decisions of Congress require the agreement of its members, who represent a variety of different constituents and interests. They may also require negotiations with the presidential administration. Therefore, the funding decisions that emerge from Congress, including any specific formulas and allocation procedures, generally represent compromises between divergent interests within a general agreement to act in a particular problem area.

Program agencies are required to administer programs within the prescribed legislation. The discretion they are given varies from program to program. Typically, many of the decisions on which estimates or quantities to use in formulas will be theirs, but the formulas themselves will not. Agencies often play an important role in formulating the rules and regulations that govern how funds are to be used by recipient jurisdictions. They have a responsibility to monitor programs and identify whether they are achieving the objectives implied in legislation.

The first-level recipients of most programs are the states and territories. They can be expected to exert pressure on program agencies to maximize their gain from formula programs. In many cases, the state government is not the ultimate spender of funds but an intermediary between the federal government and the local government. In that role, the state will sometimes have its own formula allocation program for distributing state funds to localities—counties, cities, school districts, etc. (see Chapter 7).

Two other types of organization also influence this process. National, state, and local advocacy groups with an interest in the problem being addressed by a formula allocation program will take a close interest in the operation and success of a program and may be a significant voice in promoting change or expansion of programs. Statistical agencies play a key role in providing the statistical estimates required by formulas, and in some cases in developing new data collection programs to meet the legislated needs of programs.

FEDERAL, STATE, AND INTERNATIONAL PROGRAMS

Table 1-1 provides a list of the 11 largest formula allocation programs operating in FY 1999 and 2000, with the amounts of funds obligated in each of those years. The dominant role of Medicaid in financial terms is evident. A further elaboration of a selection of key programs, with a discussion of some of their distinguishing features, is provided in Appendix B.

At the state level, large amounts of state revenues are transferred to counties, cities, school districts, and other jurisdictions through a variety of formula allocation programs. Some states have general revenue sharing programs, in which uses by the recipients are unrestricted. State programs are especially significant in the field of public education, in which state funds cover nearly half of all expenditures, compared with a less than 10 percent share provided by Title I education and other federal programs (National Research Council, 1999a:Table 2).

The use of formulas to allocate funds is not restricted to the United States, although to our knowledge no country has quite the variety of separate formula allocation programs of the United States. The federal governments of both Canada and Australia operate equalization programs for their respective provinces or states that aim to ensure that residents of poorer jurisdictions can receive common basic services at comparable rates of taxation. These programs are similar in intent to the General Revenue Sharing Program that operated in the United States between 1972 and 1986. In the United Kingdom, large amounts of money (84 billion pounds, or $126 billion, by one estimate) are distributed annually by the national government to local areas (Derbyshire, 2001). The favored approach to distributing public funds is increasingly to place more weight on objective means of allocation through mathematical formulas that model expenditure needs (Smith et al., 2001). Although some formulas are rudimentary, the general tendency has been toward ever-greater intricacy. Attempts to simplify formulas have not met with much success, as the demands for sensitivity to local conditions (and therefore complexity) seem to dominate.

International organizations face a similar allocation problem but in reverse—how to calculate the contributions that individual member countries should provide toward the organization's financing. The United Nations, for example, operates a formula-based assessment system based primarily on gross national product, but taking into account debt burden, per capita income, a ceiling for the least developed countries, and other adjustments.

The use of formulas for allocating funds is pervasive and increasing worldwide. On one hand, the appeal of a formula approach is its apparent objectivity and transparency and some separation of funding decisions from political influence once the formula has been agreed on. On the other hand, the inevitable political negotiations that must precede agreement to a formula can result in one that deviates significantly in its impact from a formula that might appear fair and rational to a disinterested bystander.

BACKGROUND, CHARGE, AND REPORT ORGANIZATION

Building on a workshop held in Spring 2000, the Committee on National Statistics (CNSTAT) formed the panel to study formulas used to allocate federal funds to states and localities. The purpose of this panel study was not to recommend changes in existing formulas, new formulas, or the use of particular data sets. Rather, through its deliberations and commissioned papers, the panel was to assess and illustrate the interplay among statistical estimates used as inputs, formula features, and the legislative process, with specific attention to attainment of program goals and the objectives of the allocation process. The charge to the panel stated:

> ...that the Committee on National Statistics will conduct a 30-month panel study of statistical issues in the allocation of federal and state program funds to states and localities. The study will consider the statistical estimates used as inputs to formulas, data and methods for estimating these inputs, the features of the formulas, and how estimates and formula features interact in ways that affect program goals. The study will be conducted in two phases. The first phase will focus on issues of interest to the U.S. Department of Education; the second phase will broaden the focus. After the first phase, the panel will issue a report that provides the panel's preliminary analyses and findings and that includes case studies of education programs as well as selected programs that provide useful points of comparison. The second phase will result in the issuance of a report that will include the panel's findings and conclusions. The panel might also issue a reference document with suggested guidelines for persons responsible for initiating, monitoring, or evaluating formula grant programs. All reports will be intended to help policy makers and others who are involved in specifying formula features and data sources for estimates for formulas.

The panel started with a primarily technical focus but soon became aware that meaningful recommendations for improvements in the structure of formulas and inputs to formula allocation processes would have to take full account of the policy and legislative contexts in which these processes are developed. The panel's thinking evolved in other ways as well. The charge to the panel made no mention of various special features of the allocation process (such as hold-harmless provisions or eligibility thresholds) and how they interact over time with basic formulas and their data inputs. However, it became evident that failure to consider such interactions can lead to unintended consequences and that this was a topic that deserved careful attention.

At the end of its first year of study, the panel issued a preliminary report of its findings and analyses, which included case studies of education programs. This final report explores the broad statistical issues, embedded in a political and policy framework, that arise in the development and use of formulas for allocating federal funds to state and local governments. The report identifies specific statistical and methodological problems associated with the use of formulas for fund allocation, such as those involving estimation, sampling and measurement error, and sensitivity analysis, but it does not seek to characterize them in detail, as most such description would necessarily be highly program specific. This type of description can be found in the literature; for example, see National Research Council (1999b and 2000) for a discussion of the statistical issues relating to Title I formula allocations.

As part of its work, the panel commissioned a set of papers that provide further support for its conclusions and recommendations by dissecting the technical underpinnings of formulas, as well as the practical application issues. The papers appear in the September 2002 issue of the *Journal of Official Statistics,* which should be viewed as a companion volume to this report. The first group of papers lays out how the formula allocation process works—the authors examine the goals and underlying structures of fund allocation formulas, describe the legislative development process, and explore how formula features, data, and estimation procedures interact in producing formula outputs. These papers are followed by several case studies that serve to illustrate many of the issues raised in the first three papers. The case studies are drawn from U.S. programs in the areas of children's health, women's and children's nutrition, and education; from a Canadian program designed to reduce discrepancies in the fiscal capacities of the provinces; and from the United Nations' dues assessment procedures.

In this final report, Chapter 2 reviews objectives of funding aid to subordinate jurisdictions, such as closing the gap between need and effort, treating economic equals equally, redistributing economic well-being, and encouraging spending on certain services. Implications of these goals for the structure of aid formulas are discussed and illustrated with some examples.

Chapters 3 through 6 provide a more detailed treatment of features that are common to most formula allocation programs, using numerous examples from U.S. federal and other aid programs. Chapter 3 outlines the broad context in which the programs operate. Who are the initial recipients and what are the program rules that govern their use and further distribution of funds? What is the timing and frequency of disbursements? What provisions are made to cover the administrative costs incurred by federal and state program agencies?

Chapter 4 identifies and provides examples of components frequently included in allocation formulas: measures of need (including geographic cost differentials), measures of the fiscal capacity of recipient governments, and measures of effort by those governments. The final section provides a model for combining those elements and discusses how errors in the input data can affect the allocations.

Chapter 5 identifies and gives examples of special features that are frequently part of the allocation process and interact with the basic formula and data inputs in determining the final results. The chapter discusses eligibility thresholds, limits, hold-harmless provisions, caps, moving averages, step functions, bonuses, and penalties.

Chapter 6 looks at sources of data for estimating formula elements. Sources are sometimes specified in legislation and sometimes determined by program agencies. The chapter discusses the advantages and disadvantages of each of the major sources used: the decennial census, current population estimates, household surveys, other statistical programs, and administrative records from sources like the Internal Revenue Service and the Food Stamp Program. Choices among these sources are influenced by both data quality and cost considerations.

In Chapters 7 and 8 the focus shifts from U.S. federal aid programs to those operated by the State of California, several foreign countries, and the United Nations. Chapter 7 begins with a brief discussion of how funds received under a number of federal formula allocation programs are managed and used by various agencies of the California state government. It then describes several programs that distribute state revenues to Californian cities, counties, and other local governments. Chapter 8 describes revenue-sharing programs in Canada, Australia, and the United Kingdom. It also describes the formula-based dues assessment system used by the United Nations to finance its operations. Although the UN system collects rather than distributes money, it is found to have many of the same features as the aid programs that are the main focus of this study.

Chapter 9 presents the panel's conclusions and recommendations.

This report has six appendices. Appendix A briefly describes the papers that were commissioned by the panel. Appendix B summarizes findings from a review of 12 federal formula allocation programs, including the 10 programs with the largest obligations for FY 1999 and two others with features of special interest. Appendix C describes the main sources of data about formula allocation programs that were used by the panel to obtain background information for this study. Appendix D outlines the contents of a proposed handbook on fund allocation formulas and processes. Appendix E lists participants in the panel's meetings and the April 2000 Workshop on Formulas for Allocating Program Funds. Appendix F provides biographical sketches of members of the panel and staff.

2

Why Provide Aid and Use Aid Formulas?

WHY HAVE AID?

In a federal system, intergovernmental aid can help policy makers accomplish many objectives. Describing these objectives is not a purely academic task, even though, as noted in Chapter 1, the political process profoundly affects the final structure of formula allocation programs. Characterizing several of the most common objectives of aid is a first step toward understanding why some aid formulas exist and how they should be structured, particularly since different aid formulas are consistent with different objectives. This section reviews four of the more common objectives of aid: closing the gap between need and effort, treating economic equals equally, redistributing economic well-being, and encouraging spending on certain services. We then discuss implications of each of these objectives for the structure of aid formulas. In practice, aid formulas often combine elements associated with more than one of the objectives described in this chapter, because the objectives are not mutually exclusive and because the designers of most aid programs hope to accomplish more than one of them.

OBJECTIVES OF AID

Closing the Gap Between Need and Effort

Many aid programs arise in response to the recognition that, if all public services provided by states or localities were financed with state-level or local-level taxes, the *effort* (as measured by the revenues raised at the state or local level) required to finance a target (or an adequate) level of services would vary dramatically. (Ladd and Yinger, 1994, refer to this concept of effort as revenue-raising capacity.) The goal of aid is to make it possible for all states or localities to provide the target level of services with a reasonable effort.

At the state level, there are numerous examples of aid programs that have been designed to accomplish this goal. For instance, most state aid to local school districts is distributed via foundation aid programs. Under such programs, the state guarantees that each school district that levies a tax rate at or above some minimum will be able to spend at least a minimum or *foundation* amount on each student. The local contribution toward this amount is first determined by applying the minimum tax rate to the tax base. The state then provides enough aid to ensure that each student generates funding equal to the foundation level or allowance.

Downes and Pogue (2002) show that foundation aid programs are consistent with the goal of enabling all aided localities to provide a target level of services with a reasonable effort. Such explicit links between the goal of closing the gap between need and effort and the design of aid formulas are not as evident for aid programs at the federal level. Nevertheless, this goal has influenced the motivation for and the design of many federal aid programs. For some programs, like the Substance Abuse and Mental Health Services Administration's Substance Abuse Prevention and Treatment Block Grant Program,[1] the statement of the program's objective makes general reference to providing each state with the ability to meet local need:

> The Substance Abuse Prevention and Treatment (SAPT) Block Grant program goal is to support substance abuse prevention and treatment programs at the State and local levels. While the SAPT Block Grant provides Federal support to addiction prevention and treatment services nationally, it empowers States to design solutions to specific addiction problems that are experienced locally (Substance Abuse and Mental Health Services Administration, 2001).

In other cases, by providing more aid to states or localities with less ability to raise revenue, formulas have the explicit intent of making more

---

[1] See Appendix B for more information about this and other formula allocation programs mentioned in this chapter.

equal the capacity of recipient jurisdictions to serve their constituents. In still other cases, federal aid formulas incorporate cost adjustments as a way of accounting for variation among recipient jurisdictions in the expenditures necessary to provide the services for which aid is being provided. Closing the gap between need and effort is therefore one of the goals, if not the sole one, motivating critical elements of many of the largest federal formula allocation programs.

In the language of economics, these gaps between need and effort are referred to as *fiscal disparities*. In the absence of aid, fiscal disparities imply that government activities, such as the provision and financing of education, are not *locationally neutral*—the taxes that individuals bear to receive a given level of public services depend on where they reside and engage in economic activities. If moving is costly, fiscal disparities mean that some individuals are worse off simply because they reside in localities with higher costs or fewer taxable resources. Closing the gap between need and effort can redistribute economic well-being to those who reside in disadvantaged communities. Furthermore, if, as seems likely, individuals who move must incur real or psychic costs, then the variation in fiscal disparities can alter the location of private-sector activities, thereby creating an inefficient pattern of economic activity. In addition, without aid, fiscal disparities can result, in those localities with limited capacity, in levels of public services that are, from a societal perspective, dangerously low.

Treating Equals Equally

Closely linked to the objective of closing the gap between need and effort is a second objective of aid: promoting the evenhanded treatment of individuals whose economic circumstances are the same. Aid is seen as a mechanism for ensuring that individuals who are alike in their ability to pay taxes or in their need for the services that aid is being used to finance do, in fact, have the same tax burdens and receive the same benefits from government services.

In the economics literature, aid that is motivated by this objective is said to promote *horizontal equity*.

Redistributing Economic Well-Being

Many aid programs arise out of a desire to redistribute resources and, ultimately, economic well-being from those with more ability to pay the taxes needed to finance publicly provided services to those with less ability to pay. Ultimately, the goal is to make access to the benefits of economic prosperity more equal.

The community development block grants offer one example of an aid program that has as its goal the redistribution of economic well-being. The grants provided under this program are intended "to develop viable urban communities, by providing decent housing and a suitable living environment, and by expanding economic opportunities, principally for persons of low and moderate income" (Appendix B). Clearly, the hope is that aid will provide low- and moderate-income individuals with improved access to the fruits of economic prosperity.

Aid programs that are designed to shift resources toward those who are less well-off economically are attempting to produce a distribution of economic well-being that is more *vertically equitable*. The degree to which the correlation between economic well-being and ability to pay is reduced determines the extent to which an aid program results in a distribution of economic well-being that is more vertically equitable.

Encourage Spending on Certain Services

Frequently, architects of aid programs design them with the hope of inducing changes in the spending behavior of recipient governments by altering the incentives they face. For example, the matching element of the State Children's Health Insurance Program (SCHIP) exists to encourage states to establish insurance programs that cover children in families who are not eligible for Medicaid and for whom private insurance is prohibitively costly (see Appendix B).

Aid can also be used to encourage recipient jurisdictions to make different uses of the resources available to them. Making aid contingent on measurable improvement in the quality of services provided, as is implicitly done in the No Child Left Behind Act of 2001, is intended to force recipient governments to eliminate inefficiencies and direct resources toward those services that are most needed by the target populations.

Aid programs can alter behavior in less subtle ways. Maintenance of effort requirements, such as those that exist under the Temporary Assistance for Needy Families (TANF) program, mandate the extent to which revenues raised by the recipient government can decline after aid has been received. Such mandates are intended to reduce the extent to which aid substitutes for local resources. Similarly, conditioning aid receipt on the implementation of specific policies, as was done when receipt of highway aid was linked to passage of state speed limit and drinking-age laws, is an example of the use of aid as a lever to get recipient governments to implement policies that federal policy makers determine to be in the national interest.

Such levers are necessary only if the national interest and the interest of the recipient government diverge. Frequently, a mismatch in local and national interests exists because the benefits of the public service in question accrue not just to the residents of the recipient jurisdiction but also to some individuals who have no voting interest in the locality's receiving aid. For example, local governments might choose to spend too little on education because the direct beneficiaries of that spending are not current voters and the community's elected leaders perceive that the lower tax rates associated with the lower spending levels will help attract commercial and industrial investment. In such situations the benefits of the public service spill over the boundaries of the providing jurisdiction either geographically or in time. If *spillovers* of this type exist, provision of the public service will be inefficient in the absence of some intervention by the state or the national government. Properly designed aid formulas can create incentives for local governments to increase or decrease spending on the aided function and thus implicitly account for the external benefits associated with that spending. In other words, linking the receipt of aid to the pursuit of certain policies can ameliorate (or even eliminate) inefficient provision of a public service attributable to a mismatch between the jurisdiction that provides the service and members of society who benefit from it.

FAIR TREATMENT OF COMMUNITIES

Objectives like closing the gap between need and effort and encouraging spending on certain services can be addressed relatively effectively using intergovernmental aid, since those objectives are stated in terms of communities, not individuals. However, the objectives of promoting equal treatment of equals and of redistributing economic well-being are stated in terms of people, not communities. Not all people in wealthy communities are wealthy; not all people in poor communities are poor. As a result, redistributing resources across communities may not result in very effective redistribution across individuals.

So, if policy makers hope to accomplish these objectives, why use intergovernmental aid, as opposed to direct aid to individuals? Intergovernmental aid programs tend to be simpler to administer than programs that provide direct aid to individuals. Furthermore, aid programs in general and formula aid programs in particular are the product of compromises between different interests. Some legislators who support a particular aid program are motivated by specific goals. For example, proponents of certain key elements of the No Child Left Behind Act, which reauthorized the Elementary and Secondary Education Act and with it the Title I education program, pushed these modifications to the Title I program because of their concern about the achievement gap between poor and disadvantaged students and their more affluent peers. Others are more concerned with the impact of these programs on the resources flowing into their states and districts. Crafting a compromise between these competing interests may be easier if aid is directed to communities rather than to individuals, since calculating the resources directed to legislators' key constituencies can be done more easily under intergovernmental aid programs and since the economic status of individuals tends to be correlated with the economic status of the communities in which they live. Those who hope to use aid to redistribute economic well-being know that poor people tend to live in poor communities. Thus, even if aid is motivated by an objective that is stated in terms of individuals, administrative simplicity or political feasibility may result in intergovernmental aid. Intergovernmental aid may, in the language of economics, be a *second-best policy*.

Nevertheless, the disconnect between individual-centered objectives of aid and the community-centered structure of aid can create confusion in the discussion of aid programs. For example, if the objective of an aid program is equal treatment of equals, those constructing and evaluating this program tend to ask if communities are treated fairly under it. But fair treatment of communities is not necessarily consistent with equal treatment of equals. Similarly, if the objective is to redistribute economic well-being, the tendency is to design an aid program to redistribute from rich to poor communities—but that may not redistribute from rich to poor individuals.

WHY USE FORMULAS?

Linking Structure of Aid to Objectives

As was noted in Chapter 1, formulas do not represent the only mechanism for allocating aid to states or localities. Do they offer any advantage over legislating shares, for example, as is done in the Evironmental Protection Agency's State Capitalization Grants Program?

One clear advantage of formulas is that they permit linking the structure of the aid program to its objectives. The commissioned paper by Downes and Pogue (2002) provides a more extensive discussion of the link between objectives and the structure of aid programs.

Suppose, for example, that the objective of aid is to make it possible for all recipient jurisdictions to provide the target level of services with a reasonable effort. In this case, the objective maps directly into the aid formula. Specifically, as Ladd and Yinger (1994) and Downes and Pogue (2002) observe, the formula that gives the appropriate amount of aid for any recipient jurisdiction is:

Aid = (Spending needed for target services) – (Local revenue raised
　　　with reasonable effort)
$$= FnC - t^*V$$

where F is the level of spending per eligible individual needed to achieve the target service level, n is the number of eligible individuals in the recipient jurisdiction, C is a cost index that adjusts for interlocality differences in the cost per eligible individual of providing given public services, t^* is the formula tax rate, which is multiplied by each recipient government's fiscal capacity to determine its contribution to financing the target level of spending, and V is the fiscal capacity of the recipient jurisdiction. The formula tax rate t^* is chosen so that, if a locality chooses to levy that rate, it will be able to provide the target service level with a reasonable effort. The formula tax rate t^* and the level of spending needed to achieve the target service level F are policy parameters; these quantities would be the same for all recipient jurisdictions. Typically, policy makers would set the value of t^* at what they feel is the minimum fair tax rate. Recipient jurisdictions typically choose local tax rates that differ from t^*; providing aid according to this formula closes the gap between need and effort but does not prevent residents of any single recipient jurisdiction from choosing to provide more or less of the public service in question.

The other objectives of aid discussed above also map directly into elements of aid formulas. For example, changing the incentives facing recipient jurisdictions can encourage spending on a particular service. This can be done using matching, which reduces the amount by which local tax revenues must be increased in order to increase spending on the aided function by one dollar. In other words, matching aid programs like SCHIP encourage more spending on the aided function by reducing the "price" of that function from the perspective of the recipient government.

Because the objectives of promoting equal treatment of equals and redistributing economic well-being focus on individuals, not communities, mapping these objectives into specific elements of aid formulas is more challenging. Still, aid can come closest to accomplishing these objectives if it is distributed according to a formula. Suppose, for example, that the goal is redistribution. Then, given that poor individuals tend to reside in poor communities, aid should go disproportionately to poor communities (Ladd, 1994). This will happen if aid is distributed using an aid formula that accounts for a community-level measure of ability to pay such as total taxable resources (TTR), with aid amounts varying inversely with ability to pay.

Facilitating Political Compromises

As was noted in Chapter 1, aid exists in a domain somewhere between the idealized world in which aid programs are designed to fulfill explicit objectives and the purely political world in which aid programs simply divvy up pots of money. Many of those who advocate for aid programs are motivated by specific objectives, but they must make compromises to get the authorizing legislation approved. Formulas can offer political cover to those involved in the process of compromising (Melnick, 2002). Also, formulas can simplify the process of compromising by reducing the dimensionality of the problem. In other words, if a formula is used to distribute aid, agreement needs to be reached on the structure of the formula and the statistical inputs to that formula. Without the aid formula, the programmatic description that allocates decision-making responsibility to a federal agency must be crafted to generate sufficient support (Melnick, 2002). Finally, formulas make it easier to quantify the impact of alternative compromises.

3

Basic Features of
Formula Allocation Programs

This chapter is an overview of the context in which federal formula allocation programs operate. Who receives the funds initially, and where do they go from there? What is the timing and frequency of disbursements to the recipients? What provisions are made to cover administrative costs? What program rules place constraints on the purposes for which the funds can be used by the recipients?

Answers to these questions vary widely by program. The way in which each program operates depends on a complex body of legislation, regulations, and policies at several levels of government. We present some examples of different arrangements, taken primarily from the 12 large federal programs whose formulas and allocation processes were systematically reviewed to provide input to the panel's deliberations (see Appendix B).

TARGET ALLOCATION UNITS

Funds appropriated for federal programs go initially to an executive branch *program agency* that is charged with using the funds in accordance with legislative requirements. For formula allocation programs, typically most of the appropriated funds are distributed by the program agency to the 50 states, the District of Columbia, and in some instances, to American Indian tribes and U.S. territories, based on formulas and procedures included in the authorizing legislation or developed by the program agency based on general objectives specified in the legislation. Funds allocated to

26

states go to various state agencies, such as state transportation agencies for highway programs and state education agencies for the special education program; these agencies use the funds for purposes defined by *program rules.*

There are some exceptions to the usual pattern of making the initial distribution to state agencies. For the Title I education program, through school year 1996-1997, funds were allocated to state education agencies and those agencies were responsible for determining the amounts to be distributed to counties and school districts. For school year 1997-1998, however, Congress decided that allocations down to the county level were to be determined by the U.S. Department of Education, and starting with school year 1999-2000 the federal responsibility for allocations was extended to the school district level. Thus, in this program, school districts are now the target allocation units.[1]

Another exception occurs in the community development block grants program, which is administered by the U.S. Department of Housing and Urban Development. Most of the appropriated funds are distributed directly to 860 metropolitan cities and 153 urban counties (counts for FY 2001) for their use in programs designed to develop viable urban communities by providing decent housing and a suitable living environment and by expanding economic opportunities, principally for people with low and moderate incomes.

FURTHER DISTRIBUTION OF FUNDS BY INITIAL RECIPIENTS

For many formula allocation programs, the ultimate beneficiaries are eligible individuals or family units who receive cash or in-kind benefits or services. Of the 12 programs reviewed in Appendix B, 7 fall into this category: Medicaid, Temporary Assistance for Needy Families (TANF), the school lunch program, special education, State Children's Health Insurance Program (SCHIP), foster care, and the Special Supplemental Nutrition Program for Women, Infants, and Children (WIC). For all of those seven except special education, eligibility is restricted to low-income families or

---

[1]An exception was made for school districts with less than 20,000 population. States are permitted to aggregate the funds allocated to such districts and to reallocate them based on alternative criteria, subject to review by the Department of Education.

individuals. To some extent, eligibility requirements are established in the authorizing legislation for the programs, but states are typically given some leeway. In the Medicaid and SCHIP programs, for example, income eligibility is determined by the ratio of family income to the official poverty level, but states are given flexibility in deciding what ratio to use and what items to count as income. The TANF program stands out as one that has given states substantial freedom to develop activities designed to meet the program's broad objectives and to establish eligibility requirements for the benefits provided.

As noted previously, in the Title I education program, funds are allocated to school districts, using formulas designed to favor districts with large numbers or proportions of school-age children in poor families. Along with much larger amounts of state aid and local funds, the Title I allotments go to support the budgets for individual schools and school district administration.

For some programs, federal funds are used to make loans to local governments for specific purposes. Under the Environmental Protection Agency's state capitalization grants program, allotments to states, along with matching state funds, are used to establish state revolving funds from which loans are made to water districts for specific waste water treatment and water pollution control projects. This system was established by 1987 amendments to the Clean Water Act; prior to that time, funds allotted to the states were disbursed as grants for eligible projects.

Highway Planning and Construction program funds are used primarily for specific projects to construct or make improvements to public (nonlocal) highways and for several related purposes. Projects proposed for federal aid funding must be included in a statewide transportation improvement program developed by the state transportation agency and submitted for approval to the Federal Highway Administration and the Federal Transit Administration. Each state's program is constrained by the funds allocated to it under the various subprograms of the Highway Planning and Construction program. For some large projects, design and construction are subject to oversight by the Federal Highway Administration.

As illustrated by these few examples (see also Chapter 7, which describes state aid programs in California), a variety of administrative mechanisms, some preexisting and some created specifically for a particular grant program, are used to channel funds from state agencies to the ultimate beneficiaries.

FREQUENCY AND TIMING OF ALLOCATIONS AND DISTRIBUTIONS

Most formula allocation programs are continuing programs supported by annual appropriations. For many of them, the authorizing legislation calls for reauthorization after a specified number of years, typically four or six. The reauthorization process gives Congress an opportunity to conduct periodic reviews of the effectiveness of the program and of the existing allocation formulas and procedures. This arrangement does not preclude more frequent changes if Congress sees a need for them. Changes in formulas, data sources, and other features of the allocation process can be enacted through special legislation, added on to other related legislation, or even enacted as part of the annual appropriations legislation for the program.

Within this general framework, requirements for sound financial management and differences between programs in the timing and predictability of funding needs have led to development of various practices with respect to frequency and timing of allocations and distributions. Some examples are:

• For Medicaid, which is an open-ended matching grant program, payments to the states are quarterly, based on statements of eligible expenses submitted by each of the state programs.

• Matching grants under the foster care program, which is also open-ended, are handled in the same way as Medicaid grants.

• In the initial stages of SCHIP, it was difficult for states that were just starting their programs to determine when those programs would be approved and what their caseloads would be. Although the program provides for federal matching of state expenditures at varying rates, the total amount to be allocated to the states each year is fixed. Consequently, the law allows states up to three years to spend their allocation for a given fiscal year. After that period, unused funds are reallocated to states whose expenditures have exceeded their allotments for that year.

• As provided for in the regulations for the WIC program (7CFR246.16), the U.S. Department of Agriculture recovers unexpended prior-year funds from the state agencies. These funds are reallocated to other states in accordance with a formula established for that purpose.

• For the Title I education program, school districts receiving funds need to have at least a rough idea of how much federal and state aid they will be receiving in order to plan for the coming school year. Allocations of

Title I funds are normally announced in the spring and are available to the school districts for obligation by the July preceding the beginning of the school year. School districts are not permitted to carry over more than 15 percent of their allotments to the next fiscal year. State education agencies have the authority to reallocate unused funds.

• In the Unemployment Insurance Program, which provides federal funds for the administration of state-financed unemployment insurance programs, the amounts required for a fiscal year depend in part on the levels of insured unemployment in each state, which are difficult to predict. The Employment and Training Administration of the U.S. Department of Labor, which administers the program, divides the total amount available for allocation to the states into base and contingency budgets. Formula-based allocations under the base budget are made available to the states at the start of each fiscal year. Contingency funds are made available to states when their quarterly workloads exceed the base budget level. This procedure makes it possible to distribute funds roughly in proportion to observed needs.

• The Canadian equalization program (discussed in Chapter 8) calculates annual provincial entitlements using a large number of data items. These allotments are preliminary estimates, which are revised twice a year until the final estimate is made 30 months after the end of the fiscal year, at which time all of the relevant data are available. Payments to the provinces are made each month, with adjustments after each revision and after the final estimate for a fiscal year.

PROVISIONS FOR ADMINISTRATIVE AND OTHER OVERHEAD COSTS

For many formula allocation programs, the authorizing legislation provides that a specified proportion of the annual appropriation shall be set aside for use by the executive branch program agency for administrative and planning purposes. In the highway program, for example, a maximum of 1.5 percent of the total appropriation is set aside for administrative activities of the federal program agencies. This is sometimes called the "administrative takedown." The legislation for the school lunch program includes two set-asides for the secretary of agriculture: a maximum of 3 percent of the appropriation for administration of the program and a maximum of 1 percent for training and studies. The substance abuse block grant program has a 5 percent set-aside for data collection and evaluation studies.

At the state level, there are various provisions for covering and in some instances capping expenditures for administration of programs whose main purpose is to pay benefits to families or individuals. A separate formula-based program, State Administrative Expenses for Child Nutrition, supports states' administrative expenditures for the school lunch program, the school breakfast program, and the special milk program. Similarly, the WIC program includes a separate formula-based grant to states for nutrition services and administration. A total of 10 percent of the grant amount for each state is withheld and pooled with amounts withheld from other states to form regional discretionary funds, which are distributed by the U.S. Department of Agriculture according to guidelines that take into account the varying needs of state agencies within each region. The unemployment insurance program, as noted earlier, exists solely to pay administrative costs of state-financed and operated programs that pay benefits to unemployed workers. The foster care program, which provides matching grants, at rates that vary by state, for direct assistance, also provides matching funds, but at uniform rates, for selected overhead costs: 75 percent for training staff and foster parents and 50 percent for administrative data collection. The SCHIP program permits states to devote up to 10 percent of their total expenditures to nonbenefit activities, such as administration, outreach, health services initiatives, and other child health assistance. These expenditures are matched for each state at its enhanced federal medical assistance percentage, the same rate used to match its expenditures for program benefits.

PROGRAM RULES

Program rules, which are established pursuant to a program's authorizing legislation, are laid out in detail in federal regulations and also in regulations and policies of state agencies or other bodies involved in the subsequent distribution of program funds and benefits. Program features covered by these rules can include eligibility requirements for individuals, families, and other program beneficiaries; acceptable uses of program funds by recipients; time limitations on the use of funds; reports by state agencies to federal program agencies; audits of program operations and expenditures; and evaluation studies. To illustrate the kinds of features that can be covered by federal program regulations, Box 3-1 shows the table of contents of

BOX 3-1
Regulations for the National School Lunch Program:
Table of Contents

SUBCHAPTER A—CHILD NUTRITION PROGRAMS

PART 210—NATIONAL SCHOOL LUNCH PROGRAM

Subpart A—General

Sec.
210.1   General purpose and scope.
210.2   Definitions.
210.3   Administration.

Subpart B—Reimbursement Process for States and School Food Authorities

210.4   Cash and donated food assistance to States.
210.5   Payment process to States.
210.6   Use of Federal funds.
210.7   Reimbursement for school food authorities.
210.8   Claims for reimbursement.

Subpart C—Requirements for School Food Authority Participation

210.9   Agreement with State agency.
210.10 What are the nutrition standards and menu planning approaches for lunches and the requirements for afterschool snacks?
210.11 Competitive food services.
210.12 Student, parent, and community involvement.
210.13 Facilities management.
210.14 Resource management.
210.15 Reporting and recordkeeping.
210.16 Food service management companies.

Subpart D—Requirements for State Agency Participation

210.17 Matching Federal funds.
210.18 Administrative reviews.
210.19 Additional responsibilities.
210.20 Reporting and recordkeeping.

Subpart E—State Agency and School Food Authority Responsibilities

210.21 Procurement.
210.22 Audits.
210.23 Other responsibilities.

Subpart F—Additional Provisions

210.24 Withholding payments.
210.25 Suspension, termination, and grant closeout procedures.
210.26 Penalties.
210.27 Educational prohibitions.
210.28 Pilot project exemptions.
210.29 Management evaluations.
210.30 Regional office addresses.
210.31 OMB control numbers.

APPENDIX A TO PART 210. ALTERNATE FOODS FOR MEALS
APPENDIX B TO PART 210. CATEGORIES OF FOODS OF
   MINIMAL NUTRITIONAL VALUE
APPENDIX C TO PART 210. CHILD NUTRITION LABELING
   PROGRAM

AUTHORITY: 42 U.S.C. 1751Ø1760, 1779.

Source: 53 FR 29147, Aug. 2, 1988, unless otherwise noted.

the current regulations for the National School Lunch Program. The authorizing legislation for this program stresses the importance of providing nutritionally adequate lunches. The regulations include a section (210.10) on nutrition standards and three lengthy appendices with detailed provisions on the use of meat protein substitutes, a listing of foods of minimum nutritional value, and a description of food labeling procedures.

Program rules are extremely varied in their content and level of detail. A detailed analysis would go well beyond the scope of this study. To some extent, as noted in Chapter 6 under "Bonuses and Penalties," the rules can affect the amounts received by the target allocation units.

In the April 2000 workshop that preceded and motivated the formation of this panel, one participant made a distinction between formula allocation procedures, which determine how much money goes to each recipient jurisdiction, and program rules, which determine, with varying degrees of specificity, how the money is to be used by the recipients. This participant argued that marginal adjustments in the distribution formulas were considerably less influential in determining the degree to which program goals are met than are the rules governing what happens after the funds are received. The panel agrees that program rules are a major determinant of success in achieving program goals, but in our view allocation formulas and procedures are important and deserve close attention. The issues of potentially unnecessary formula complexity and the interaction between formula features and inputs require additional study.

4

Components of Allocation Formulas

As their name implies, formula allocation programs allocate funds to recipients (governments or individuals) via a formula. Many formulas depend on the need of the recipient, and it is also common for formulas to include measures of fiscal capacity. And it is not unusual for aid amounts to depend on local effort. These and other inputs are common components in many formulas; however, the manner in which they are operationalized and combined does differ. Appendix B, a review of 12 of the largest federal formula allocation programs explores these issues in relation to specific programs.

MEASURES OF NEED

In theory, the resources needed by a recipient government in order to provide a target level of services depend both on the number of individuals eligible for the services and on the cost of providing them to each eligible individual. Generally, both of these components of need must be estimated. For example, the goal of the State Children's Health Insurance Program (SCHIP) is to encourage states to establish insurance programs that cover children in families who are not eligible for Medicaid and for whom private insurance is prohibitively costly. Thus, aid under SCHIP depends on the estimated number of children who are not eligible for Medicaid and who do not have private insurance. Importantly, the cost of providing health care to each uninsured child varies by state and, if possible, this cross-state variation should be reflected in aid amounts. But the true cost of insuring each eligible child will not be known and must be estimated.

Many aid formulas include ad hoc adjustments for variation in need. For example, to compensate for geographic differences in prevailing salaries and thereby in the cost of education, the Title I, Part A, grants to local education agencies formula includes state per-pupil expenditure (PPE) as one factor in determining the allocation (Brown, 2002). However, to our knowledge no existing school aid formula includes cost adjustments that are closely linked to evidence on the costs of providing particular services. There is general agreement that aid formulas should account for differences in costs, but generally they do not because developing cost estimates is extremely difficult and somewhat contentious. For example, although differences in PPE reflect relative costs, they also reflect variations in local wealth and commitment to education.

The SCHIP and Title I education examples highlight the important issue of whether aid should compensate primarily for costs that are beyond the control of decision makers in the recipient jurisdictions. If compensation is also influenced by controllable costs, the aid formula could generate perverse or undesirable incentives. For example, in SCHIP the method used to estimate the number of eligible children must be designed to avoid penalizing states for the success of the program (Czajka and Jabine, 2002). Also, if administrative data are used to estimate the number of eligible children or the cost per eligible child, these data must be chosen so that they are immune to manipulation.

A critical task for researchers is to compare the results of the alternative methodologies with the goal of developing consensus estimates of need. Duncombe and Lukemeyer (2002) provide a superb model of the style of research that must be produced.

MEASURES OF FISCAL CAPACITY

Measures of fiscal capacity are the second shared component of aid formulas. While per capita personal income is the most commonly used measure of fiscal capacity, other measures are more consistent with the goals of individual aid programs (Tannenwald, 1999; Downes and Pogue, 2002). For example, Downes and Pogue (1994) and Ladd (1994) agree that, if the goal of the aid program is to close the gap between need and effort and thereby reduce fiscal disparities, capacity measures based on the representative tax system (RTS), variants of which are used in state-level school aid programs and in the Canadian equalization program, should be used. The use of RTS-style measures of capacity ensures that all localities receiving aid will be able to finance a basic package of public services by levying average tax rates. Thus, no recipient governments will be at a competitive disadvantage (Ladd, 1994).

If the goal is redistribution of economic well-being or equal treatment of economic equals, the appropriate choice is a capacity measure that is based on an adjusted measure of income, for example, the total taxable resources (TTR) measure of capacity that is used in the community mental health services block grant and substance abuse block grant programs. TTR is based on the rationale that if the objective of aid is to redistribute from rich to poor or is to promote the evenhanded treatment of individuals whose economic circumstances are the same, the measure of capacity should reflect the cumulative ability of the residents of a locality receiving aid to pay taxes (Tannenwald, 1999). Using such a measure of capacity ensures that localities with equal abilities to pay taxes for public services will be treated equally and that localities with less ability to pay will get more aid. TTR is preferred to per capita income because it provides a more comprehensive measure of ability to pay. For example, accrued capital gains are included in TTR.

Any measure of capacity must be estimated, usually at a refined level of geographic and temporal resolution. The difficulties of generating income estimates for small areas are well known (National Research Council, 2001). The estimation problems become more daunting when adjustments need to be made for the extent to which taxes can be exported (shifted to non-residents) (Bradbury and Ladd, 1985; Downes and Pogue, 1994). Even capacity measures derived using the RTS methodology can be subject to large errors. Taylor et al. (2002) document some of the measurement problems that arise in the Canadian context; Tannenwald (1999, 2001) provides graphic evidence of the difficulty of generating state-level measures of fiscal capacity using the RTS methodology.

MEASURING EFFORT

A measure of effort is the final component common to many aid formulas. Recall that effort is measured by the revenues raised at the state or local level and spent on providing the service for which aid is provided. Effort measures are central elements in matching aid formulas, since for most recipient jurisdictions the amount of aid is a (possibly variable) percentage of effort. Effort is also an important part of any aid formula that imposes a minimum effort requirement.

It would seem that effort is the component of an aid formula that is easiest to measure using direct information. Local tax rates are readily observable, as is spending on the service for which aid is being provided. However, measuring effort is not quite so straightforward. If aid amounts depend on spending on the targeted services, the recipient government has a strong incentive to overstate effort by classifying spending on other services as spending on the targeted services.[1] Similar incentives to overstate effort operate if aid is contingent on minimum effort. Nevertheless, of the three shared components of aid formulas, effort can be measured with the smallest error.

COMBINING COMPONENTS

Most formulas combine two or more measures of need, fiscal capacity, and effort. Appendix B provides examples of the varied ways in which these measures are combined. Ideally, for programs with explicit goals, components should be combined to target these goals (Downes and Pogue, 2002); however, frequently there is not such a tight link. No matter how a formula is constructed, the choice of a particular combination of the components will influence how they interact.

A simple example helps clarify how errors in the measurement or estimation of the different components of an aid formula can interact. Chapter 2 presents a formula showing how components might be combined to close the gap between need and effort.

Errors in the number of eligibles (n), the level of spending per eligible individual needed to achieve the target service level (F), the cost index that adjusts for interlocality differences in the cost per eligible individual of providing given public services (C), or the fiscal capacity of the recipient jurisdiction (V) will produce discrepancies between the actual and desired aid distributions.

Reducing the error in measuring one component may not produce an improvement. Suppose, for example, that fiscal capacity is measured with-

---

[1]In the rare instances in which aid amounts are reduced by increased local effort, the recipient governments might attempt to understate effort by hiding expenditures or using funding mechanisms that are not counted as part of local effort.

out error but that the errors in estimating the number of eligibles and the cost per eligible are negatively correlated. The aid distributed might then be relatively close to the desired distribution. Reducing the error in estimating the number of eligibles while making no improvement in estimating the cost per eligible could result in an aid distribution that lines up less well with the desired distribution.

Changes in the aid distribution resulting from proposed improvements in the measurement of one or more components should be evaluated. Evaluations should include a study of how the relations between formula outputs (allocations) and inputs (measures of need, fiscal capacity, and effort) are affected by the change, taking into account the effects of hold-harmless provisions and other special features of the allocation process.

5

Special Features of Formula Allocations

For many programs, a basic formula is not in itself sufficient to meet all of the objectives of the allocation process. Special features are introduced for various reasons: to promote more efficient use of program funds, to reflect fixed costs of program operations, to avoid disruptions caused by large year-to-year changes in amounts received, or to negotiate passage and authorization of the allocation program. Generally, a basic formula is a continuous function of such inputs as need, fiscal capacity, and effort; special features can introduce discontinuities. They can sometimes result in a grant system in which resources are not well matched with needs as they change over time.

Special features, such as thresholds for eligibility and upper or lower limits on total amounts (or on match rates or components of the basic allocation formula), operate on an individual year basis. Other features, such as hold-harmless provisions or caps, place constraints on the year-to-year change in the amounts allocated to each recipient jurisdiction. This chapter gives examples of these special features, discusses possible rationales for their inclusion, and examines some of their impacts and consequences.

THRESHOLDS

Thresholds on eligibility to receive funds produce discontinuities. The Title I education program, the community development block grant program, and the HIV Emergency Relief Project Grants under Title I of the

Ryan White Care Act impose such thresholds. In the Title I education program, individual school districts are grant recipients. To be eligible for a *basic grant*, a school district must have at least 10 eligible children *and* a poverty rate for school-age children of 2 percent or more. A likely rationale for the minimum number of eligibles is that a critical mass of funding is needed to enable a program; funding below that level would be too small to run a Title I program of sufficient size, scope, quality, and impact. A possible rationale for the minimum rate requirement is that a wealthy school district with no more than a 2 percent poverty rate would have relatively little need for Title I funding—such a district would have sufficient resources to address the needs of its at-risk students. Title I *concentration grants* are designed to provide funds for areas with large concentrations of poor families: a school district must have at least 6,500 eligible children *or* a 15 percent poverty rate. For either type of grant, a shortfall of one eligible child means that no grant funds are received.

Eligibility for these two grants is based on estimates of the numbers of eligible children and poverty rates. These estimates are based on data from sample surveys, the decennial census, and administrative sources and are subject to substantial statistical variation, especially for the smaller school districts. Some school districts that would be eligible if exact counts were available receive no funds, and some that would not be eligible do receive funds. These false negatives can have a large impact on both large and small school districts. A small school district with 10 truly eligible children that failed to receive a basic grant because the estimate was 9 or fewer might be unable to serve these children, and the amount lost could be substantial for a large school district with a true poverty rate at or slightly above the 2 percent level or one whose number of eligible children was close to the 6,500 threshold for concentration grants. In a capped program, false positives take funds away from truly qualifying districts.

The community development block grant program features another type of threshold. Annual appropriations for the program are divided, with 70 percent directly allocated by the U.S. Department of Housing and Urban Development to eligible metropolitan cities and urban counties (entitlement communities) and 30 percent allocated to states to be distributed to nonentitled jurisdictions that must apply to the state for funding. Entitlement communities include central cities of metropolitan statistical areas (MSAs), other cities with at least 50,000 population in MSAs, and urban counties, which are counties located in MSAs and having a population of at least 200,000, excluding entitled cities. Decennial census counts or intercensal population estimates are the basis for determining which cities and counties have attained these threshold sizes.

Thresholds established for grants under Title I of the Ryan White Care Act are based on population and the number of reported AIDS cases. A metropolitan area becomes eligible for emergency relief project grants if it has a population of 500,000 or more and has reported a cumulative total of more than 2,000 cases of AIDS for the most recent 5 calendar years for which data are available from the Centers for Disease Control and Prevention. Once eligibility is established, the area remains eligible.

Zaslavsky and Schirm (2002) conducted simulations to assess how an eligibility threshold in a formula interacts with sampling error in the measure of eligibility that is used. They found that as sampling error increases, the sharp cutoff that is seemingly implied by the threshold is replaced by an increasingly smooth relation between an area's true need and its expected allocation. The reason is that, as sampling variability increases, an area with true need below the threshold is more likely to have estimated need above the threshold, while an area with true need above the threshold is more likely to have estimated need below the threshold. On average, areas with true need below the threshold get more than they deserve, while areas with true need above the threshold get less than they deserve. Allocations for small areas, which typically have smaller samples and larger sampling errors, tend to be distorted more than allocations for larger areas. Importantly, allocations depend both on formula features and on the statistical properties of estimated formula inputs, which in turn depend on the design and analysis of sample surveys and other inputs. This interaction between formula features and statistical properties of inputs has high leverage and apparently has seldom been taken into consideration by those who design formulas and surveys. Its single-year (cross-sectional) and year-to-year (longitudinal) impacts need to be evaluated in detail.

LIMITS

Upper and lower limits are used in various ways to constrain the outcomes that would result if an allocation were determined solely by a basic formula. In some instances, limits are imposed on the values that can be taken by selected components of the basic formula. Alternatively, limits may be placed on the allocation determined by the basic formula, whether it be expressed as an amount, a share of the total appropriation, or a federal matching proportion.

Examples of such limits can be found in the Title I education and the substance abuse block grant programs. In Title I, state per pupil expenditure (which is multiplied by the estimated number of eligible children to provide an estimate of need) is restricted to lie between 80 and 120 percent of the national average per pupil expenditure. In the block grant program, the cost of services index for a state must lie between 0.9 and 1.1.

In financial terms, the most significant limits are those placed on the federal medical assistance percentage (FMAP), which is used in Medicaid and several other formula allocation programs to determine what proportion of state program expenditures will be reimbursed by the federal government. The formula for FMAP is:

$$FMAP = 1.00 - 0.45 \times \left[\frac{StatePCI}{NationalPCI} \right]^2$$

with the constraint that

$$0.50 \leq FMAP \leq 0.83$$

The lower limit of 50 percent was retained from a predecessor program that provided a flat matching rate of 50 percent to all states (U.S. General Accounting Office, 1983).

At present, no states have per capita income so low that they are affected by the 0.83 upper limit. However, several states with high per capita incomes (11 in FY 2002) receive 50 percent matching funds, which is more than they would receive if there were no lower limit. An unconstrained FMAP matching percentage could be derived from a theoretical construct of how to level the playing field. Therefore, limits on the matching percentage may indicate disagreement on such a construct or other, possibly political, considerations. For some programs, limits are relatively easy to justify. For example, in the federal aid highway program, no state can receive less than 90.5 percent of its estimated contributions to the Highway Trust Fund, which funds the program. The Highway Trust Fund is financed by receipts from user taxes, and it is reasonable that each state should receive at least some minimum proportion of the taxes it provides. In the special education program, no state may receive more than an amount equal to the number of its children receiving special education services multiplied by

40 percent of the average per pupil expenditure in U.S. public elementary and secondary schools. The long-range goal established for this program is to provide federal funding for 40 percent of the cost of special education in each state, but so far the annual appropriations have been insufficient to reach this level of support.

Limits can be used to dampen the effects of outliers for formula inputs based on estimates that are highly variable. Limits can also be used to preclude attempts by recipients to increase their receipts by manipulating formula inputs.

Several formula allocation programs place a lower limit on the proportion of the total funds allocated to be received by any state. The federal aid highway and EPA state capitalization grants programs each guarantee a minimum share of 0.5 percent to every state, and the substance abuse block grants program guarantees a minimum of 0.375 percent. We have identified two possible justifications for such small-state minimums, one practical and one political. Most programs require states to incur expenses to set up programs to administer the receipt and use of federal grant funds, and some of the costs may be more or less fixed regardless of a state's population. The other consideration is that all states, regardless of population, have two senators and their votes are needed to pass authorization and appropriation legislation.

HOLD-HARMLESS PROVISIONS AND CAPS

Limits are imposed on year-to-year changes in the amounts received by states or other recipients. Hold-harmless provisions limit downside changes; caps limit increases.

Big swings in allocations can be of great concern to legislators and administrators. Legislators for areas whose amounts or shares decline are likely to face difficult questions from their constituents. On one hand, unpredictable declines in federal program funding can cause difficulties for state and local program administrators, for example, school officials planning budgets for the coming year. On the other hand, most fund allocation programs are designed to meet specific needs and to equalize, at least in part, the fiscal capacity to meet those needs. As needs and fiscal capacities change, allocations should be responsive to those changes. Therefore, except for open-ended programs like Medicaid and foster care, there is a clear trade-off and tension between stability and addressing current needs, especially if annual program funding remains level or declines.

Improving formula inputs may improve targeting, primarily by updating and upgrading the quality of the data used to estimate needs and fiscal capacity. Sometimes improved estimators, such as the model-based estimates that have been developed for the Title I education and WIC programs, can be developed and introduced into the formula allocation process. Although on the average they may reflect needs more accurately than the inputs previously used, they are still subject to statistical error, which can be relatively large. For example, the Title I education program requires school-district-specific estimates, and the State Children's Health Insurance Program (SCHIP) requires state estimates for a narrowly defined subset of the total population. These estimates usually have a high relative variance.

Hold-Harmless Provisions

In order to maintain some degree of stability, several programs include *hold-harmless provisions* that guarantee that each recipient entity will receive, at a minimum, a specified proportion of the prior year's amount[1] or share. The specified proportion may be 100 percent, or it may be less than 100 percent. In some programs, hold-harmless provisions remain the same from year to year; in others they are in force for a limited period, especially at times when revised formulas were being introduced.

Hold-harmless provisions, in combination with year-to-year changes in total funding, limit the extent to which allocations reflect changes in need. In an extreme case, if there is 100 percent hold harmless and no increase in funding, there will be no change from the previous year's allocations, regardless of any changes in need or other elements of the basic allocation formula.

In at least two programs, hold-harmless provisions have largely neutralized efforts to improve the targeting of allocations to current needs. In the Title I education program, a model-based estimation procedure, with estimates updated biennially, has replaced the earlier use of estimates based on the decennial census, which were updated only every 10 years. However, for FY 1998-2001 Congress enacted a 100 percent hold-harmless provision and provided only a modest increase in the annual appropriation. Therefore, the revised estimates had little effect in shifting funds to the

---

[1]In some instances, the total amount available may be insufficient to meet the hold-harmless guarantee; in such cases, allocations to all participants are "ratably reduced" so that their sum is equal to the total funds available.

areas where needs had increased relatively the most rapidly. Indeed, there were instances in which school districts that qualified for concentration grants for the first time received no funds because there was nothing left over after the hold-harmless provision had been satisfied for school districts that had received grants in the previous year.[2]

In the WIC program, in which similar steps have been taken to improve estimates of need, there is a 100 percent hold-harmless provision (called a "stability grant"). Of the funds available for food grants after allowing for stability grants, 80 percent is used to cover increases in food costs due to inflation, and only the remaining 20 percent is allocated to states that are not receiving their fair shares as determined from current estimates of need.

The special education program allocation rules ensure that as long as there is an increase in funds compared with the preceding fiscal year, no state can receive less than its allocation for that year (a conditional 100 percent hold harmless). Additional provisions ensure that states will receive some minimum proportion of any increase in the amount appropriated for the current fiscal year. SCHIP has a hold-harmless provision that applies to shares rather than amounts. Starting with FY 2000, no state's share can be less than 90 percent of its share for the preceding fiscal year or less than 70 percent of its FY 1999 share.

Zaslavsky and Schirm (2002) conducted simulation studies of the relations between hold-harmless provisions and variability in estimates of formula inputs. They found that when there is a hold-harmless provision in a formula that allows an area's allocation to rise by any amount but fall by only a limited amount, sampling variability in estimates of formula inputs ratchets up allocations over time. Such ratcheting occurs because sampling variability can raise an area's allocation—perhaps substantially— in a year, but the hold-harmless provision always prevents the allocation from falling very much the next year. The amount of ratcheting increases as sampling variability increases. Because estimates for smaller areas typically have greater sampling variability than estimates for larger areas, the upward bias in allocations from hold harmless is greater for smaller areas; thus, the smaller areas tend to benefit more from a hold-harmless provision than do larger areas.

---

[2]In school year 2000-2001, 23 newly eligible school districts received no concentration grant allocations (see Brown, 2002).

Caps

Caps, which limit the size of increases, are less common than hold-harmless provisions. In the special education program, no state can receive more than its allocation for the previous year increased by the percentage increase in the total amount appropriated plus 1.5 percent. In SCHIP, starting in FY 2000, no state's share can exceed 145 percent of its share for FY 1999.

MOVING AVERAGES AND HOLD-HARMLESS

Stabilizing formula inputs, for example by using moving averages computed by averaging estimates from two or more consecutive years, will stabilize formula outputs. For example, the FMAP matching percentage used in Medicaid and several other matching grant programs specifies the use of moving averages of estimates of state and national per capita income for the three most recent years.

As discussed before, Zaslavsky and Schirm (2002) found that sampling variability in estimates of formula inputs ratchets up allocations over time when there is a hold-harmless provision. They also found that using moving average estimates can greatly reduce the biasing effect of a hold-harmless provision and, in fact, can be as effective or more effective than a hold-harmless provision in moderating downward fluctuations in funding. Although moving average estimates will tend to be too high if there is a downward trend in need for an area and too low if there is an upward trend in need, Zaslavsky and Schirm show that using an exponentially weighted moving average that gives more weight to more current data reduces this tendency for a moving average to lag behind a trend.

Although the use of moving averages in allocation formulas can eliminate or sufficiently reduce ratcheting, it may not in itself be sufficient to achieve the desired level of stability, and hold harmless may still be needed. The authorizing legislation for SCHIP required that the allocation formula use three-year moving averages of Current Population Survey state estimates of children eligible for the program. However, the year-to-year variation in these moving averages proved to be so large that, as noted earlier, it was considered necessary to introduce a hold-harmless provision ensuring that no state would receive less than 90 percent of its share for the preceding year.

STEP FUNCTIONS

The statutory hold-harmless provision for basic grants under the Title I education program specifies the guaranteed proportion of the prior year's grant as a function of the estimated poverty rate (the percentage of eligible children in the school-age population). The hold-harmless proportion is 95 percent if the estimated poverty rate exceeds 30 percent, 90 percent if the estimate is between 15 and 30 percent, and 85 percent if the estimate is below 15 percent. Thus, a difference of one person in the numerator or denominator of the estimated poverty rate could make a substantial difference in the amount received by a school district. Such discontinuities can be avoided by making the hold-harmless proportion a smooth and slowly changing function of the poverty rate.[3]

BONUSES AND PENALTIES

Several formula allocation programs have provisions for bonuses and penalties. Although these provisions usually do not affect the initial allocations for the current fiscal year, they can lead to subsequent additions to or subtractions from a state's allocation as determined by the basic formula. The Temporary Assistance for Needy Families (TANF) program includes an annual appropriation of $1 billion through FY 2003 for bonuses to the states that do best in moving aid recipients into jobs and an appropriation of $100 million to reward states that are most successful in reducing the number of out-of-wedlock births and abortions. The program also has contingency and loan funds that are used to assist states experiencing economic downturns.

Penalty provisions are more common than bonuses. TANF has several penalties that are assessed on states failing to satisfy work requirements, failing to comply with paternity establishment and child support enforcement requirements, and failing to meet state maintenance of effort requirements. Penalties are assessed as a varying percentage of state allocations and states that are penalized must expend additional state funds to replace the amounts lost. The total of penalties assessed in a fiscal year may not exceed 25 percent of a state's block grant.

---

[3]In recent years, the statutory provision has been overridden by insertion at the appropriation stage of a 100 percent hold-harmless provision for both basic and concentration grants.

The federal aid highway program, which includes highway planning and construction and several other subprograms, has numerous penalty provisions. State allotments for various subprograms can be reduced by specified percentages for failure to enforce vehicle size and weight laws, to control outdoor advertising, to comply with the 1990 Clean Air Act Amendments, to have a law that prohibits purchase or public possession of any alcoholic beverage by a person under age 21, and to comply with other requirements.

The national school lunch program includes maintenance of effort requirements. States are required to appropriate or use for program purposes state funds amounting to up to 30 percent of the federal funds received. The 30 percent requirement is reduced for states whose per capita income is less than the national average. If a state fails to meet its requirement, certain federal funds are subject to recall by or repayment to the U.S. Food and Nutrition Service.

These examples illustrate the carrot and stick aspect of federal grant programs, whereby payments to states are made contingent on behavior considered by Congress and executive branch agencies to be in the national interest.[4]

SUMMARY

Special features, such as thresholds, limits, hold-harmless provisions, and caps, are used in many formula allocation programs to serve various purposes. They can lead to allocations quite different from those that would result if only the basic formula is used. Any attempt to evaluate the performance of a formula allocation program must, of necessity, take account of the effects of these special features, both on initial allocations and on changes over time.

Generally, policy analysts and congressional staff simulate one year's allocations under different formula provisions, and sometimes longitudinal assessments are conducted to explore how allocations vary over time under different hold-harmless levels (including no hold harmless) and other provisions. Similarly, longitudinal analyses are needed to assess how program operations in an area might be affected by, for example, a 10 percent reduction in funding, especially if need had really decreased by 10 percent.

---

[4]For more on this topic, see the series of articles on welfare quality control programs (estimation of penalties), from the *Journal of the American Statistical Association*, vol. 85, no. 411 (1990): articles by Kramer (pages 850-855), Hansen and Tepping (856-863), Fairley et al. (874-890), Puma and Hoaglin (891-899).

6

Data Sources for Estimating
Formula Components

Components of allocation formulas fall into three broad categories: need, fiscal capacity, and effort. These components were defined and several examples given in Chapter 4; here we consider the sources of data used to estimate these components. Specifically, we discuss how data sources for each formula component are determined, which sources are most commonly used, and what considerations are relevant in choosing from among alternative data sources.

WHO DETERMINES WHAT DATA SOURCES ARE TO BE USED?

For some programs, the original authorizing legislation or various amendments to it are very specific about the form of the allocation formula and data sources for each formula element. The current legislative requirement for the State Children's Health Insurance Program (SCHIP), for example, specifies that the estimate of the number of uninsured children for each state is to be based on the three most recent March supplements to the Census Bureau's Current Population Survey (CPS) before the beginning of the calendar year in which the fiscal year begins. For other programs, Congress delegates authority for such decisions to the program agency or the secretary of the department in which the agency resides. For example, a provision of the Individuals with Disabilities Education Act, the authorizing legislation for the special education program, states that: "For the purpose of making grants under this paragraph, the Secretary [of

Education] shall use the most recent population data, including data on children living in poverty, that are available and satisfactory to the Secretary" (20 U.S.C., Ch. 33, Subchapter II, Section 1411(a)).

Even when a data source is clearly specified in legislation (such as the CPS in the SCHIP example) the actual estimates from that source will be affected by survey design and procedures. These factors are largely determined by the program agency and the agency responsible for collecting the data, subject to budgetary constraints. For example, to reduce the high sampling variability of CPS state-specific estimates of uninsured low-income children, starting with FY 2000 Congress provided an annual appropriation of $10 million to increase the sizes of the relevant samples.

For the Title I education program, Congress has given the U.S. Department of Education considerable flexibility in deciding on data sources. Prior to the mid-1990s, estimates of the number of poor school-age children by state were required by law to be taken from the most recent decennial census.[1] By the end of each decade, these estimates could be seriously deficient as a basis for estimating the current distribution of need. The Improving America's Schools Act, passed in 1994, called for the use of updated Census Bureau estimates of poor school-age children to allocate Title I funds, provided the estimates were found to be sufficiently reliable by a panel of the National Research Council (NRC). In response to the 1994 act, the Census Bureau established a small-area income and poverty estimates (SAIPE) program to develop estimates by state, county, and ultimately by school district, using a model-based approach that combined data from the decennial census, the CPS, and administrative records. With review and eventual advice from the NRC Panel on Estimates of Poverty for Small Geographic Areas, these estimates were adopted for use in allocating Title I funds (National Research Council, 2000).

As noted in Chapter 1, for the Special Supplemental Nutrition Program for Women, Infants, and Children (WIC) program Congress defined the program objectives in legislation and left it to the agency to develop the allocation formula. Thus, the Food and Nutrition Service of the U.S. Department of Agriculture was responsible for determining what data sources and procedures should be used to estimate formula components.

---

[1]For fiscal years 1979 to 1987, one-half of the excess over the fiscal year 1979 appropriation was allocated to states on the basis of data from the 1976 Survey of Income and Education.

Since the allocation formula was first used in 1979, the program agency has made several improvements in the estimates (National Research Council, 2001:32-34).

DATA SOURCES

A wide variety of data sources are used to estimate formula components for the more than 180 federal formula allocation programs. These sources are grouped below into four major categories.

Decennial Census and Current Population Estimates

Every 10 years, the decennial census provides counts of population by age, sex, race, and Hispanic origin for states, counties, cities, and other political subdivisions such as school districts. For censuses through 2000, a long-form sample has provided additional data, at the same level of geographic detail, for persons, families, and housing units. These decennial census sample data, in particular information on income, have been widely used as inputs to allocation formulas. The Census Bureau has recently announced plans to eliminate the long-form sample from the 2010 decennial census. A large-scale continuing household survey, the American Community Survey (ACS), is intended replace it, producing continuously updated data of similar content. When cumulated over 5-year periods, the ACS data will provide estimates of roughly the same precision as the decennial long form.

For several decades the Census Bureau has provided intercensal estimates of population, using data on births, deaths, immigration, and internal migration to "walk" the estimates from the most recent census to the current date. Expansion of this current estimates and projections program has been driven to a substantial degree by the requirements of formula allocation programs. From 1972 through 1986, estimates of population and per capita income for approximately 39,000 units of local government were required for allocations under the general revenue sharing program. Current population estimates by state serve as the denominator for the estimates of per capita income used in the formula for the federal matching assistance percentage (FMAP), used in Medicaid and several other formula allocation programs to determine federal matching rates by state. More recently, the SAIPE program was established to produce current estimates of school-age children in poverty by county and school district for use in the Title I education allocations.

A study by the U.S. General Accounting Office (1990) determined that nearly two-thirds of federal formula funds were distributed either wholly or in part using population data from the decennial census. Due to their lack of timeliness and the availability of more current estimates from the Census Bureau, decennial census data are now used less as formula inputs.[2] The community development block grants program provides an exception; several of the elements in the two alternative formulas use data from the most recent decennial census. Also, Census data are required for metropolitan cities and urban counties for such characteristics as population in poverty, the number of housing units with more than 1.01 persons per room, and the number of housing units built before 1940. Such data have not been readily available from any current household surveys; however, a fully operational ACS should provide more current data for some of these variables.

Household Surveys

The CPS provides monthly estimates of employment and unemployment for various population subgroups. Its annual March supplement provides data on individual and family income for the preceding calendar year. CPS income data are an important input to the SAIPE model-based estimate of school-age children in poverty that are used in the Title I education allocations. Data from the March supplement are also used in model-based estimates of infants and children eligible for the WIC program by state, as well as in state estimates of uninsured and total low-income children for SCHIP.

The National Household Survey of Drug Abuse, which was recently expanded to provide direct estimates for eight states and synthetic estimates for the remaining states, has the potential to provide better estimates of need than those currently used in the allocation formula for the substance abuse block grant program. However, like most household surveys, it does not cover institutionalized populations.

---

[2]In a 1999 study, the U.S. General Accounting Office developed estimates of the effects of using census data adjusted for estimated undercount on FY 1998 allocations by state in 25 large formula grant programs (U.S. General Accounting Office, 1999).

Other Statistical Programs

As part of its system of national income and product accounts, the Bureau of Economic Analysis publishes annual estimates of per capita personal income for the nation, for states, and for selected metropolitan areas and counties. These estimates are based on a combination of survey and administrative data. The state estimates, averaged over the latest three-year period, are used as a proxy measure of relative fiscal capacity in the FMAP formula. Thus, this data source affects distributions in programs that account for more than half of the total federal funds allocated each year. The Treasury Department's series on total taxable resources, which is based on data from the Bureau of Economic Analysis and the Internal Revenue Service's Statistics of Income Division, is used as a measure of state fiscal capacity in the substance abuse and mental health block grant programs.

Wage and price statistics from several sources are used in some programs to account for geographic differences in the cost of program services. The cost of services index in the formulas for the substance abuse and mental health block grant programs uses data on manufacturing wages from the Bureau of Labor Statistics' Current Employment Statistics Survey and data on fair-market rents from the U.S. Department of Housing and Urban Development. The food grant portion of the school lunch program uses data from the food away from home component of the Bureau of Labor Statistics' consumer price index for annual updates of national average prices for free, reduced price, and paid lunches. The cost factor in the allocation formula for SCHIP is based on Bureau of Labor Statistics data on mean annual wages in the health services industry.

Allocations in the Environmental Protection Agency's (EPA) Clean Water State Revolving Fund are based on information about infrastructure needs of public water systems identified in periodic drinking water needs surveys. Prior to fiscal year 1988, allocations in EPA's state capitalization grants program were based in part on specific needs for waste water treatment and water pollution control, as identified in the periodic clean water needs surveys.

Administrative and Program Records

Administrative and program records play a major role in the determination of amounts allocated to states and other recipients in many programs. For open-ended matching grant programs such as Medicare and foster care, formula-based matching proportions are applied to state records of eligible program expenditures to determine the amounts to which the states are entitled. For the school lunch program's food grants, allocations to states are based on their records of the number of paid, reduced price, and free lunches served. In the initial years of the special education program, allocations to states depended on the number of children participating in their programs. Data on expenditures and enrollment in elementary and secondary public schools, collected from the states by the National Center for Education Statistics, are used to calculate the state per pupil expenditure component of the Title I education allocation formula. State-provided data on vehicle miles traveled on the interstate system, lane miles and vehicle miles traveled on principal arterial routes (excluding the interstate system), and diesel fuel used on highways are inputs to allocation formulas used for subprograms of the federal aid highway program.

Model-based estimates of need, which combine data from several different sources, have made substantial use of administrative data. Data on tax returns from the Internal Revenue Service and on participation in the food stamp program from the U.S. Department of Agriculture have been used to estimate need components in allocation formulas for the Title I education and WIC programs. WIC has also made use of data on unemployment insurance claims.

CONSIDERATIONS IN THE SELECTION OF DATA SOURCES

As previously indicated, data sources to be used in estimating formula inputs are sometimes specified in the authorizing legislation; sometimes the choice is left to the program agency. In all situations, factors to consider in deciding what data sources to use, relate to data quality and to evaluation of the costs and benefits associated with the use of alternative data sources.

Formula allocation programs provide for annual allocations for a specified or indefinite number of years. Choices of data sources for the estimation of formula inputs can be influenced by the initial allocations and by the way allocations change from year to year. Choices may be influenced by how program designers evaluate trade-offs between relative stability in annual funding and responsiveness to changes in the distribution of true need among recipients.

Data Quality Considerations

Data from a census, a survey, or other statistical or administrative record source have several attributes that may be relevant to their suitability for use in formula allocations:

- The *conceptual fit* between currently available data and the formula elements, as defined in authorizing legislation or administrative regulations. If the definitions of the elements or program goals lack specificity, evaluation of fit may require subjective judgments. Even if the primary goal of program designers is to arrive at a predetermined allocation, choice of data sources that provide a good conceptual fit may improve the initial and ongoing credibility of the allocation process.
- The *level of geographic detail* at which data are provided. Most programs allocate appropriated funds to the state level; a few allocate funds to smaller areas, such as metropolitan areas, counties, or school districts. The decennial census can provide estimates for areas as small as school districts (although with substantial sampling variability for the smaller districts), whereas the Survey of Income and Program Participation (believed to provide more precise estimates of family income) can provide reasonably stable direct estimates for only a few large states.
- The *timeliness* of the data, the elapsed time between the reference period for the estimates and the period for which the allocations are being made. Late in a decade, decennial census data are at an obvious disadvantage compared with continuing or periodic sample surveys and administrative record sources.
- The levels of *sampling variability* and *bias* associated with the data. It is important that these factors be evaluated in terms of their expected effects both on initial distributions and on year-to-year changes in allocations.
- The *susceptibility of the data to manipulation* by program recipients. Of necessity, such data as state program expenditures in matching grant programs must be generated by recipients of the grant funds. In such circumstances, to ensure accurate reporting, program agencies issue regulations that define standard concepts and definitions for use in reporting and develop quality control and audit procedures.

There are many trade-offs among these quality considerations, and it is unlikely that any single data source will be uniformly superior to others.

Trade-offs can be illustrated by comparing alternative sources of income data. At the national level, the most comprehensive, individual-level data on income by source come from the Survey of Income and Program Participation, but it has the smallest sample size. CPS data are somewhat less detailed but are based on a larger sample. Individual income tax data are not subject to sampling error but cover only about 90 percent of the total population and are based on income concepts that differ from those used in most fund allocation programs. However, their utility would be improved if they were geocoded to the county and school district levels.[3] Decennial census data cover a larger proportion of the population and provide more geographic detail, but they lack timeliness and are subject to greater underreporting of some types of income. Model-based estimates that combine information from these data sources have the potential to take advantage of the strengths of each source. The SAIPE estimates developed by the Census Bureau for use in the Title I education formula program provide an excellent example of the modeling approach.

Cost-Benefit Considerations

Obtaining data to be used as inputs to allocation formulas is by no means cost free. Even when data sources created for other purposes are used, there may be significant costs of obtaining data in a suitable format, mapping variable definitions into those needed and evaluating the performance of the inputs. Hence, the potential benefits conferred by improving conceptual fit or other aspects of data quality have to be weighed against the cost of such improvements. Spencer (1985) discusses some ways that a cost-benefit analysis for a statistical data program can take into account the trade-offs between nonoptimality of an allocation formula and the cost of improving it.

If a formula is designed to produce a predetermined allocation, one may question the need for high-quality statistical data to serve as inputs to the formula. However, if a formula is designed to meet specified program goals, it is possible that social welfare is increased when high-quality data are used—for instance, high-quality data may help maintain support for a program if it conveys the sense that the allocations are fair and responsive. But these benefits must be assessed, especially when a closed-ended formula

---

[3]State income tax data for some states are available at this level of detail.

is used to distribute a fixed total, so that underallocation to some areas is matched algebraically to overallocations to other areas. Relevant assessments include simulations to determine how the use of higher quality data is likely to change the allocations. Determining the likely program effects of such changes is much more difficult. The sometimes tenuous links between improved inputs to formulas and attainment of goals for increases in social welfare have implications for how much to invest in developing new or improved data for allocation purposes. If the link is strong, then the benefit from spending more money to improve the statistics will be more than if the link is weak.

7

A State View—California

In a study of formula allocation programs, states are of interest both as recipients of federal funds and as distributors of their own funds to cities, counties, and other jurisdictions. In order to obtain an admittedly limited but nevertheless useful view of these state roles, the panel focused on the State of California. The first section of this chapter explains how California distributes federal program funds, using examples from 7 of the top 11 federal programs (see Table 1-1). The second section describes methods used to allocate funds in selected state-funded aid programs.

ALLOCATION OF FEDERALLY PROVIDED FUNDS

Distribution of federal funds within a state is, to a large degree, controlled by the federal program agencies. In some programs, funds flow to the state and are disbursed through the state agency that administers the program. Sometimes a single federal program may be administered by more than one branch of a state agency or by more than one state agency. In a few programs, for example Title I education, the federal government allocates funds directly to substate entities, with only minor discretion allowed.

• The federal *Medical Assistance Program (Medicaid)* is implemented through state regulations and administered by the California Department of Health Services. There are no issues of suballocating funds within the state. The state is billed by and reimburses providers for services to eligible recipients. States are reimbursed 75 percent of the program's administrative costs, and there is a 50 percent match of the state's general fund expenditures to pay service providers. The program has no cap or ceiling. Administrators of state Medicaid programs are in continuous communication through various stakeholder advisory groups, such as the Medicaid Management Information System and associated subcommittees. This interaction fosters information exchange and the identification of best practices for program administration.

• *Federal-Aid Highway* funds are administered by the California Department of Transportation. Working with local governments, the state has responsibility for identifying and prioritizing projects with a minimum horizon of three years for inclusion in a state transportation improvement program document. There is an iterative process between the state and regional planning organizations for projects in urbanized areas. Federal funds generally represent 80 percent of the project cost, although the percentage is higher for some projects, such as traffic signalization. The Federal Assistance Award Data System third-quarter report for 2000 shows nearly 800 action records (one for each project) in California. The largest was $30 million for widening a San Jose highway. All state departments of transportation belong to the American Association of State Highway and Transportation Officials. This group, collaborating through working groups and committees, influences national standards and policies and recommends positions on legislation.

• *Temporary Assistance for Needy Families (TANF)* is provided through California Work Opportunity and Responsibility to Kids (CalWORKs). The state plan, designed to help CalWORKs recipients find employment and/or acquire job skills necessary to obtain employment, was developed in consultation with local governments and private-sector organizations and approved by the secretary of the U.S. Department of Health and Human Services. The program is supervised by the California Department of Social Services and administered by county welfare departments. The departments of social services, child support services, and health services, and the Office of Criminal Justice Planning have authority to make rules and regulations that ensure universal access and uniform eligibility criteria for the program. Other state agencies, such as the Department of Education, the Employment Development Department, and the Community Colleges Chancellor's Office, are involved in education and employment aspects of the program.

The Department of Social Services distributes funds to county welfare departments for the activities associated with providing benefit payments, required work activities, and supportive services. Over half of the fiscal year CalWORKs funds are distributed in a single allocation check that provides counties with the flexibility to use funds interchangeably for eligible recipients. There are four components in the $1.7 billion single allocation for FY 2001-2002: eligibility administration (22 percent), welfare-to-work employment services (42 percent), Cal Learn assistance for teen parents (2 percent), and child care support (34 percent).

Stage One Child Care funds, representing 34 percent of the single allocation for child care support, cover care during the period beginning with a family's entry into the CalWORKs program until six months later, or when the financial situation is stable. Almost $575 million was allocated in 2000-2001 based on the following formula:

— Thirty percent of the funds was distributed based on a percentage of total payments during the most recent three state fiscal year quarters (June 2000 through March 2001).

— Seventy percent was allocated using each county's number of children through age 12 on aid multiplied by the cost per child in child care for that county.

In FY 2001-2002, child care funds were reduced by 1.74 percent, and each county's allocation was reduced by the same percentage. The counties were guaranteed their adjusted actual expenditures for the most recent quarters. Child care reserve funds may also be available to meet child care needs that exceed the counties' allocations.

- *Title I of the Elementary and Secondary Education Act* serves approximately 4,800 schools (58 percent) in 833 (83 percent) of California's school districts. Funds received by the districts are allocated to their member schools based on the number of low-income students according to school lunch program data on free or reduced price meals and/or CalWORKS data. The state's school districts receive a total annual appropriation of almost $1 billion (State Controller's Office, State of California, 2002).

Prior to 1999-2000, the State of California Title I Office received the federal funds and allocated them among school districts, based on the number of disadvantaged students, estimated from approvals to participate in free and reduced price meal programs. It was possible for a school district to receive fewer funds compared with a previous year if growth in the number of poor students in other districts was greater. Once a school district received its allocation, the district decided how to distribute funds among its schools. A district could decide to allocate the limited Title I funds to schools with the highest percentage of disadvantaged students, allocating no funds to schools with a low percentage of eligible students.

• *Special Education, Grants to States,* providing federal funds through the Individuals with Disabilities Education Act to support the expense of educating students with disabilities, is administered by the California Department of Education. States are required to allocate funds to Special Education Local Plan Areas (SELPAs), employing the same formula used to determine awards to the states. The formula has three components: the base amount, the population amount, and the poverty amount. States must uniformly apply the best available data on the numbers of children enrolled in public and private elementary and secondary schools and the numbers of children living in poverty. The California Department of Education uses data from its enrollment accounting system for the population amount. The poverty amount is based on free and reduced price meal participation data collected by CalWORKS.

The California Budget Act further establishes three separate program grants based on age and institutional grouping to each SELPAs subgrant. The local assistance entitlements grant is allocated for students 5 to 21 years old. The state institutions grant supports the special education needs of those in state special schools, California Youth Authority facilities, and Department of Developmental Services facilities. The preschool local entitlements grant provides funding for special education and services to preschool children aged 3 through 5 with disabilities.

State program directors receive information, technical assistance, and communication support from the U.S. Department of Education and the National Association of the State Directors of Special Education. The department maintains regional technical assistance resource centers in addition to funding several centers for the study of topical areas in special education.

The maximum amount that can be held for administration varies by state, and states have a fair amount of latitude in the use of program funds in the area of administration. Some states, like California, allocate funds to local education areas that could be held for administration. California has over 100 SELPAs defined by demographic and geographic criteria.

- California's *State Children's Health Insurance Program (SCHIP)* is a state and federally funded health coverage program for children with family incomes above the level eligible for no cost Medi-Cal and below 250 percent of the federal poverty income guidelines. For California, 65 percent of the program funds are federal and 35 percent are state. The state's Healthy Families Program currently serves more than half a million children. The governor and the secretary of the U.S. Department of Health and Human Services recently announced approval of the state's request for a waiver to expand the program's coverage to custodial parents, legal guardians, and family caregivers. This could increase SCHIP coverage to an additional 300,000 low-income families. States have the option to provide coverage by expanding their Medicaid programs, establishing separate new programs, or a combination of the two. California opted to use the combined approach.

Federally organized SCHIP technical advisory groups and the National Academy for State Health Policy disseminate program information and provide state participants with conferencing and communication opportunities.

- *Community Development Block Grants (CDBG), Nonentitlement Grants* are distributed by the U.S. Department of Housing and Urban Development (HUD) to participating states based on a statutory formula that includes population, poverty, overcrowded housing, and age of housing. States must ensure that at least 70 percent of the grant funds are used for activities that benefit low- and moderate-income persons. At least 51 percent of the grant funds must be used for housing.

The California Department of Housing and Community Development administers this program to approximately 180 small cities, rural counties, nonrecognized Indian tribes, and *rancherias* (rural American Indian settlements) in the state. Eligible localities include cities with populations less than 50,000 and counties with populations less than 200,000 that do not receive CDBG funds directly from HUD. The current $46.1 million grant to California is disbursed to eligible jurisdictions through several allocations: general, American Indian, *colonias* (distressed nonentitlement jurisdictions within 150 miles of the California-Mexico border), economic development, and planning and technical assistance.

State regulations mandate certain set-aside amounts: 5 percent for activities benefiting the residents of *colonias*, 1.25 percent for grants on behalf of nonrecognized tribes and *rancherias*, and 30 percent for economic development activities. Regulations also mandate how the economic development dollars are to be allocated: 10 percent are made available for planning and technical assistance on a continuous basis. Funding decisions are based on a first-come, first-served basis, with the application of eligibility threshold criteria. Separate funds available for the California Community Economic Enterprise Fund component are used to provide capital or infrastructure assistance to businesses that create private-sector jobs for low- and very-low-income persons. The allocation is based on prior years' demand and is announced annually.

The general allocation, the largest component at $25 million, receives the funds remaining after the mandated set-asides are satisfied and contains two funding mechanisms. As in the economic development component, 10 percent is set aside for planning and technical assistance grants, also on a continuing, first-come, first-served basis. Remaining funds are allocated through a once-yearly process of application review for a broad variety of CDBG-eligible activities, such as housing rehabilitation, public facilities, new construction, community facilities, public services, and economic development activities.

STATE-FUNDED PROGRAMS

California counties and school districts receive substantial revenues from state and federal government agencies. In FY 1998-1999 the counties reported revenues of $38.4 billion. The state was the primary external source of revenue, providing 31 percent of their financing. Federal agency contributions represented an additional 19 percent of total county finances (State Controller's Office, State of California, 2001).

The state was the largest contributor to public school districts. Of the schools' FY 1997-1998 revenues of $35.4 billion, 53 percent were provided from state aid, state lottery, and other state funds. Local revenues (31 percent), other financing sources (8 percent), and the federal government (slightly less than 8 percent) supplied the remaining school funding (State Controller's Office, State of California, 2000b).

The state, counties, and school districts receive a large share of revenues from intergovernmental sources. Some of these transactions are based on formula allocations. The State Controller's Office accounts for and controls disbursement of all state funds and issues warrants in payment of the state's bills, including lottery prizes. The Controller's Office annually disburses formula-based payments for 36 programs, including 4 payments that allocate federal dollars. We describe several programs that highlight the use of formulas in the allocation of California's state aid funds.

- *Monthly Motor Vehicle License Fees ($3.6 billion in FY 2000-2001).* The motor vehicle license fee (VLF) is a fee on the ownership of a registered vehicle in California in lieu of a personal property tax on vehicles. It represents a discretionary revenue source and a major source of funding for state and local health and welfare programs. The formula for allocating the VLF funds is:

> — 81.25 percent is paid to cities and county unincorporated areas based on population.
> — 18.75 percent is paid to counties.

Each year $180.9 million is paid based on revenue received in the 1982-1983 fiscal year pursuant to former Government Code sections. The amount remaining is paid to counties based on population.

- *Monthly Health and Welfare Realignment Allocation ($3.2 billion in FY 2000-2001).* The Health and Welfare Realignment Program was created in the early 1990s to transfer financial responsibility from the state to local governments for many public and mental health programs; there have been a couple of amendments since that time. Funding is provided to all counties and four cities through a portion of the state's sales tax and motor vehicle license fees.

Funds are deposited to the local revenue fund, which has accounts for sales tax, vehicle license fee, vehicle license collection, sales tax growth, and vehicle license fee growth. The sales tax account has subaccounts for mental health, social services, and health. The sales tax growth account has subaccounts titled caseload, base restoration, indigent health equity, community health equity, mental health equity, state hospital mental health equity, county medical services, general growth, and special equity.

The amount allocated for each year becomes the base amount for the following year. Revenue is multiplied by the county ratio, specified in code, for each of the subaccounts listed above. The subaccount amounts are summed to determine the total allocation amount. Any additional revenue is considered growth and is allocated on the basis of different formulas specified in legislation.

• *Monthly Half-Cent Sales Tax for Public Safety ($2.3 billion in FY 2000-2001).* Funds are allocated to counties for public safety services that include, but are not limited to, sheriffs, police, fire protection, county district attorneys, county corrections, and ocean lifeguards. Public safety services do not include courts. Payments are made to each county in proportion to its share of the total taxable sales in all qualified counties during the most recent calendar year for which the State Board of Equalization has reported sales.

• *Quarterly Lottery Apportionments ($1.1 billion in FY 2000-2001).* The voters of California passed the Lottery Initiative (Proposition 37) in the November 1984 election to provide additional funds for public education in grades K-14 and higher education. The Lottery Act mandates that public education must receive at least 34 percent of the sales revenues each year. The schools also receive unclaimed prize money, interest income, and any administrative savings at the end of each year. Lottery revenues are disbursed quarterly through counties to school districts based on the number of full-time students, measured by average daily attendance, enrolled in each district.

• *Monthly Highway Users Tax ($1.035 billion in FY 2000-2001).* Monies in the highway users tax account in the transportation tax fund are appropriated for research, planning, construction, improvement, maintenance, and operation of public streets and highways; research and planning for exclusive public mass transit guideways; construction and improvement of exclusive public mass transit guideways; and payment of principal and interest on voter-approved bonds issued for these purposes.

Several California Streets and Highways Code sections designate basic formulas to be used to allocate funds to counties and cities. Some funds are paid only to counties based on six separate calculations. In addition to a fixed amount for engineering and administration expense and a portion of the annual amounts available for snow removal and heavy rainfall and storm damage, three formulas allocate funds based on the county percentage of registered vehicles and maintained miles of roads.

Some funds are paid to counties and cities. The county payment amount is determined by calculating the greater factor of

— the legislated amount of $1,000,000 times the county's percentage of prior year's amounts from other sections of the law or

— the legislated amount of $750,000 times the county percentage of registered vehicles plus the legislated amount of $250,000 times the county percentage of maintained mileage.

The greater factor for each county is divided by the sum of all counties' greatest amounts and multiplied by a specified apportionment amount. The city payment amount is the apportionment amount divided by the total population of the state times the population of each city.

Additional sections provide different fixed amounts to counties and cities with the balance of funds distributed to counties based on percentage of registered vehicles' assessed valuation outside the cities. Cities receive other funds based either on population or a fixed amount plus an amount based on population.

- *Traffic Congestion Relief Program ($400 million in FY 2001-2002)*. Monies to relieve traffic congestion are allocated 50 percent to cities and 50 percent to counties. The 50 percent distributed to counties is based

— 75 percent on the county proportion of the number of fee-paid and exempt vehicles registered in the state.

— 25 percent on the county proportion of miles of maintained county roads in the state.

The 50 percent disbursed to cities is in proportion to the city population compared with the total population of all the cities in the state.

- *Homeowners' Property Tax Relief ($398.4 million in FY 2000-2001)*. The state constitution grants a homeowner's property tax exemption that results in revenue loss to the counties. Counties submit revenue losses to the state and payment is prorated to all counties based on the proportion of county property tax exemptions to total property tax exemptions.

- *Citizens' Option for Public Safety (COPS) Program ($242.6 million in FY 2000-2001)*. The COPS program is a public safety program to hire more sworn peace officers at the local level. Payment to counties is based on population with a minimum $100,000 grant for each recipient. The county auditor allocates monies and interest with the requirement that they be used as follows:

— 5.15 percent for county jail construction and operation.

— 5.15 percent to the district attorney for criminal prosecution.

— 39.70 percent to the county and the cities within the county; in the case of five counties with districts that encompass more than one county, specific districts are named and allocations are determined either in accordance with the relative population of the cities within the county and the unincorporated area of the county, as specified in the most recent January estimate of the population, and as adjusted to provide a grant of at least $100,000 to each law enforcement jurisdiction.

— 50 percent to the county to implement a comprehensive multi-agency juvenile justice plan.

• *Local Government Fiscal Relief ($212 million in FY 2000-2001).* In fiscal year 2000-2001, $212 million was distributed for a one-time discretionary relief to local governments. The allocation formula was:

— $100 million among cities, counties, and special districts according to their relative 1999-2000 educational revenue augmentation fund (ERAF) contributions.

— $100 million among cities based on population and counties based on population in unincorporated areas.

— $10 million to counties based on total county population.

— $2 million to independent recreation and park districts and independent library districts based on their relative ERAF contributions.

• *Quarterly State Transit Assistance ($115.8 million in FY 2000-2001).* The transit assistance funds are paid to transit districts and commissions. Half are allocated on the basis of population and half are allocated based on prior year revenues of the transit district in relation to all districts in the state.

• *California Law Enforcement Equipment Program High Technology Grants ($75 million in FY 2000-2001).* Funds are intended for grants to local enforcement agencies for the purchase of high-technology equipment. Each city, county, and specified district is guaranteed a minimum allocation amount. For 2000-2001 the minimum was $100,000; for 2001-2002, the minimum was $30,000. The balance of any remaining funds will be allocated to county sheriffs and local police chiefs in accordance with the proportion of the state's total population that resides in each county and city based on the most recent January population estimate developed by the Department of Finance.

   • *Revenue and Taxation Code Section 11005.7 ($50 million in FY 2000-2001).* This section provided $25 million to cities and $25 million to counties based on population. This is a source of general revenue to local governments. More than 80 percent of the state's population resides in incorporated cities.

CONCLUDING OBSERVATIONS

   The California experience shows that many of the principles and practices associated with allocation of federal funds by formula apply at the state level. When there is discretion in determining within-state distribution of federal funds, the state agencies develop guidelines and procedures. For several of the federal programs, state program administrators have formed organizations that enable them to share information about their procedures and experiences and to provide feedback to the federal program agencies and, in some instances, to advocate changes in legislation or regulations. California allocates substantial state funds through a wide variety of formula-based programs. Formulas are similar to those used at the federal level with some modifications and innovations to address local conditions. Some aspects of particular interest are:

   • Estimates of total population by area play a somewhat larger role than they do in federal programs. State funds are distributed based on current estimates of population by county and city provided annually by the California Department of Finance. The state prepares independent population estimates with the view that they are more timely and accurate than those produced by the federal government.
   • When data on school-age children in low-income families are needed, California uses counts of students approved to receive free and reduced-price lunches served under the National School Lunch Program. As at the federal level, there is at least one state program in which the shares are specified in legislation (see the section on Local Government Fiscal Relief above).

- Unlike the federal government, (since the end of the General Revenue Sharing program) California has several state aid programs for which uses of funds by the recipients are totally discretionary or are restricted to very broad program areas.

8

International Perspective

Looking at some experiences with formula allocation programs outside the United States, the panel examined two such programs in detail: the United Nations' (UN) assessment scale methodology for allocating the expenses of the organization among member states (Suzara, 2002), and Canada's Equalization Program (Taylor et al., 2002). We also comment on experiences in Australia and some European countries.

In general, the panel found that the statistical issues and problems arising in these programs were similar to those in U.S. programs. The choice of how to measure the fiscal capacity, or capacity to pay, of jurisdictions arises in every program. Many programs put caps or ceilings on the amounts (or the changes in amounts) that can be allocated to jurisdictions, in order to constrain expenditure for the financing jurisdiction and to preserve stability for the recipient jurisdiction. The issue of whether and how to account for differential costs of providing services across jurisdictions arises in most programs. Some of the ways in which these statistical issues have been addressed by other countries and international organizations may provide guidance to U.S. programs.

However, the different political and administrative contexts that operate in other jurisdictions make these aspects of foreign programs less applicable to U.S. programs. The ways in which programs are developed, approved, and administered depend to such a large extent on the political and cultural environment in which they operate that some practices may not be easily adapted to the situation in this country. Nevertheless, we point out some practices that are likely to be beneficial if they could be adapted.

UNITED NATIONS

Unlike the other formula allocation programs the panel studied, the UN's system for determining member states' contributions involves a formula that allocates a tax rather than a benefit. Nevertheless, it has many of the characteristics of conventional formula allocation programs that transfer funds in the opposite direction. When the United Nations came into existence it instituted a mechanism that would systematically allocate the contributions to be secured from member states to finance its operations. Thus a scale of assessments was formulated and since then has been the basis on which the expenses are distributed among the membership. It was recognized that no perfect formulation existed or could exist. Instead it was understood that the best formula was one for which a consensus existed. The underlying principle of this scale of assessment was that expenses should be apportioned according to capacity to pay. At that time it was recognized that it would be difficult to measure such capacity merely by statistical means and that it would be impossible to arrive at any definitive formula. In order to avoid anomalous assessments resulting from the use of comparative estimates of national income, other factors were also to be considered. Taken together, all of these have led to the basic elements of the current assessments methodology.

• Income as measure of capacity to pay. The economic basis for assessment is the concept of "capacity to pay." Comparative estimates of income (gross national product or GNP) are determined to be the fairest guide in measuring this capacity. Other measures, such as wealth, socioeconomic indicators, dependence on one or a few primary products, and deteriorating terms of trade, had also been considered, but problems arising with the availability, reliability, and comparability of existing data for all member states precluded their use.

• Low per capita income allowance. In order to prevent anomalous assessments resulting from the use of comparative levels of income, comparative income per head of population is factored into the formula through the application of the low per capita income allowance formula (LPAF). LPAF embodies the principle that citizens of a rich country contribute a larger share of their taxes toward the United Nations than those of a poor country who need to allocate a larger part of their income to basic necessities. The LPAF derives a common yardstick called "assessable income" that reduces the assessable income of members with large populations by the percentage difference between per capita income and a per capita income threshold corresponding to the average per capita income of all UN members. Countries having per capita income equal to or greater than the income threshold absorb the relief obtained from this application.

- Maximum and minimum rates of assessment. These rates are considered to be constraints in the scale that have been accepted as inherently political decisions and not in strict conformity with the principle of capacity to pay. At the same time, it is recognized to be an essential mechanism in an organization in which there is a wide divergence in the range of income levels of members. The maximum or ceiling rate was instituted as a means of reducing the financial dependence of the organization on a single member without seriously obscuring the relation between a member's contribution and its capacity to pay. The minimum or floor rate is based on the premise that the collective financial responsibility of an organization is borne by the entire membership and that each member pays at least a minimum fee in order to belong.

- A per capita ceiling principle specifies that the per capita contribution of any member state not exceed that of the highest paying contributor.

- An allowance to ease the burden of heavily indebted member states who devote a large portion of foreign earnings toward the servicing of external debt.

- A cap of 0.01 percent of total expenditures on the assessment rates of the least developed countries.

- A scheme of limits designed to mitigate extreme variations in assessments between two successive scales.

- A mitigation process whereby the resulting scale derived from the step-by-step application of the methodology is adjusted in order to take account of relevant factors, such as natural disasters and civil strife, that could have possible impact on capacity to pay.

The first three elements have always been part of the scale of assessments while the other elements have been included as part of the methodology at one time or another.

CANADIAN EQUALIZATION PROGRAM

In Canada, the concept of equalization dates back to the country's foundation in 1867 and was incorporated into the Constitution Act of 1982. The Canadian Equalization Program aims to ensure that "provincial governments have sufficient revenues to provide reasonably comparable levels of public service at reasonably comparable levels of taxation" (Taylor et al., 2002). It is part of a broad system of federal-provincial fiscal arrangements, which includes many other provincial and territorial transfers. The program uses the concept of per capita fiscal capacity as the basis for payments. Each province's ability to generate revenues, measured by applying national average tax rates to commonly defined provincial tax bases, is compared on a per capita basis with a common standard. Provinces below the standard receive their shortfall in per capita fiscal capacity multiplied by their population; provinces above the standard receive nothing (and contribute nothing). The common standard is currently defined as the average per capita fiscal capacity for five "middling" provinces (the four Atlantic provinces, which are relatively poor, and Alberta, which is relatively rich, are excluded). Figures 8-1, 8-2, and 8-3 show the results of this calculation for 2001-2002 and illustrate how poorer provinces are brought up to the standard, leaving richer provinces above the standard.

In some senses, the program is not unlike the general revenue sharing program that operated in the United States in the 1970s and 1980s. The equalization payments are not tied to any particular expenditures; the provinces are free to use them as they see fit. So, while the program ensures that funding is available to provide comparable levels of service, it does not guarantee that any particular service is provided.

Some features of the Canadian system may represent practice that could be emulated in some U.S. programs:

• An overall budgetary cap on total equalization payments, if it kicks in, causes the same per capita reduction in payments in each recipient province (equivalent to lowering the per capita standard), rather than a proportionate reduction of the entitlement of each province. This favors the poorest provinces.

• Almost all statistical estimates that go into the formulas, many of them coming from provincial governments, are assembled and checked by Statistics Canada (Canada's central statistical agency), which certifies them and passes them to the Department of Finance.

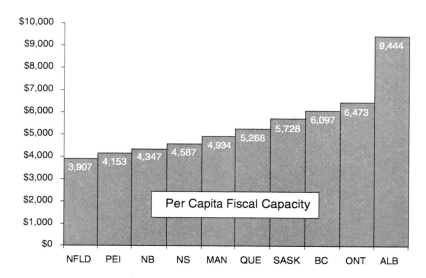

FIGURE 8-1 Per capita fiscal capacity 2001-2002 (Canadian dollars).
Note: The abbreviations on the horizontal line of the figure represent the 10 Cana-
dian provinces. They are: NFLD = Newfoundland, PEI = Prince Edward Island,
NB = New Brunswick, NS = Nova Scotia, MAN = Manitoba, QUE = Quebec,
SASK = Saskatchewan, BC = British Columbia, ONT = Ontario, and
ALB = Alberta (Taylor et al., 2002).

• Special measures have been put in place to counter distorting
impacts of sudden price shifts in some commodities and dominance of a
particular source of revenue by one province.
• Initial payments are made on the basis of estimated figures and
adjusted as final data become available. This helps to provide early warn-
ings of changes in equalization payments and provide time to soften their
impact. In addition, a hold-harmless provision limits year-to-year decline
in the per capita entitlement of any province to 1.6 percent of the standard.

AUSTRALIA, THE UNITED KINGDOM, AND
THE EUROPEAN UNION

Australia also has a long history of federal assistance to its poorer states.
It has operated an equalization program for its states since the 1970s. The

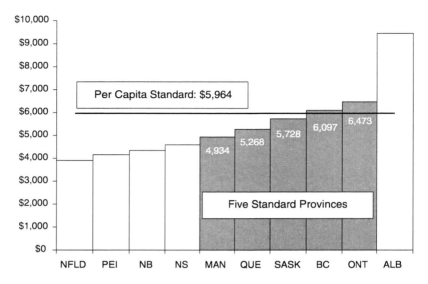

FIGURE 8-2 The per capita standard 2001-2002 (Canadian dollars).
Note: The abbreviations on the horizontal line of the figure represent the 10 Canadian provinces. They are: NFLD = Newfoundland, PEI = Prince Edward Island, NB = New Brunswick, NS = Nova Scotia, MAN = Manitoba, QUE = Quebec, SASK = Saskatchewan, BC = British Columbia, ONT = Ontario, and ALB = Alberta (Taylor et al., 2002).

Commonwealth Grants Commission, established by an Act of Parliament in 1933, provides independent advice to the Australian Parliament on intergovernmental financial relations, particularly grants of financial assistance by the commonwealth (i.e., federal) government to the states.

In the United Kingdom, also, large sums of money are distributed as a result of formula allocations. A recent study by Smith et al. (2001) noted that "the favored approach to distributing large amounts of public funds is increasingly to place weight on more 'objective' means of allocation which entails the use of mathematical formulae to model expenditure needs" (p. 218). They also noted a tendency for these formulas to become increasingly intricate. They noted that attempts to simplify formulas often fail, as the distributor of funds comes under increasing pressure to make the system more sensitive to local concerns. They report that this issue has also been examined by a House of Commons committee, which concluded that "fairness and equity are more important than simplicity, and that the drive

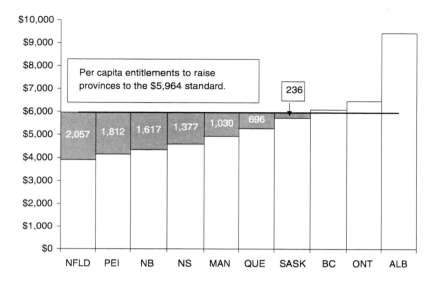

FIGURE 8-3 Per capita equalization entitlements 2000-2001 (Canadian dollars). Note: The abbreviations on the horizontal line of the figure represent the 10 Canadian provinces. They are: NFLD = Newfoundland, PEI = Prince Edward Island, NB = New Brunswick, NS = Nova Scotia, MAN = Manitoba, QUE = Quebec, SASK = Saskatchewan, BC = British Columbia, ONT = Ontario, and ALB = Alberta (Taylor et al., 2002).

to improve transparency should be a secondary concern" (Smith et al., 2001:218).

The formulas used in the financing of the European Union's budget are also complicated, as might be expected for a supranational organization.[1] Beginning in 1970, a system of national contributions was replaced by a system more like tax revenues comprising three sources:

- Agricultural levies.
- Customs duties, derived from the application of the common customs tariff to the customs value of goods imported from nonmember countries.

---

[1]Additional detail is available from the web site of the European Union: http://www.europa.eu.

• Income from the value added tax (VAT), derived from the application of a uniform rate to the VAT assessment base of each member state, harmonized in accordance with agreed-upon rules.

By 1984 these resources had become insufficient and in 1988 the European Community created an additional category of revenue, based on the gross domestic product of each member state. At that time a revenue ceiling of 1.2 percent of total European Community GNP was set. This was subsequently raised to 1.21 percent in 1995 and ultimately to 1.2⁻ percent in 1999. At the same time, the VAT rate was gradually reduced from 1.4 percent to 1 percent in 1999, and the VAT base was also cut.

In addition to following a principle of capacity to pay, the European Community has also shown concerns with equity and attempts to rationalize anomalous situations. The former is illustrated by the fact that the budget of the European Union has a cohesion fund to reinforce equity and that those eligible for assistance under the cohesion fund had their VAT base capped at 50 percent of GNP in 1995. As noted earlier, the UN assessment scale provides for relief to nations affected by anomalous situations, such as a sudden and temporary severe distortion of the foreign exchange values of their currencies. Such situations have also arisen in the European Community. In one instance, it was agreed by the member states of the European Union that the financing of one of the members needed a specific correction. In 2000 this decision was confirmed and, moreover, new rules were established with respect to the correction, with the share of the financing of the correction of some of the counties going down and a ceiling on the amount of the correction.

9

Conclusions and Recommendations

Mathematical formulas were used to allocate more than $250 billion of federal funds in 2000 to state and local governments via approximately 180 grant-in-aid programs designed to meet a wide spectrum of economic and social objectives. Large amounts of state revenues are also distributed through formula allocation programs to counties, cities, and other jurisdictions. The essential feature of a formula allocation program is that the amounts to be allocated are determined by a formula that uses statistical information to calculate or estimate the values of its inputs, and these are processed to produce outputs. Often, the allocation process consists of a basic calculation using a mathematical formula or algorithm, followed by adjustments that place constraints on levels or shares (percentages of the total allocation) or on changes in levels or shares.

In addition to providing a mechanism for addressing changes in need and other formula components without the need for Congress to revisit the issue annually, formula-based allocations can help build consensus and credibility by:

- Creating a transparent means of allocating funds.
- Creating a relatively solid foundation on which to negotiate legislation.
- Separating the question of how to distribute aid funds from the question of why they are needed.

- Creating the appearance, if not always the reality, of a sound analytic process.
- Providing a starting point for the reauthorization process.

ROLES OF CONGRESS AND THE PROGRAM AGENCIES

Program agencies in the executive branch vary in the latitude Congress gives them to decide on the basic formula to be used, the variables (e.g., population size, tax revenue, per capita income) to be used to represent formula components, and the statistical data series to be used to calculate or estimate their values. Variants differ in the control retained by Congress and the required congressional technical expertise and monitoring:[1]

a. Congress specifies program goals and intentions, leaving the program agency to specify the allocation formula.
b. Congress specifies the formula, and the program agency specifies which variables should be used and which statistics should be used to estimate them.
c. Congress specifies the formula and variables, and the program agency specifies which statistics are to be used to estimate the variables.
d. Congress specifies the formula, the variables to be used, and the statistics to be used to estimate the variables.
e. Congress specifies the numerical allocations without the need for an explicit formula.

Variant (a) allows Congress to delegate all technical issues—to build consensus on a program without resolving the fine details. If the specification of goals and intentions is sufficiently precise, relevant program agencies can use their technical expertise to develop the details. However, if goals are vague or the agency has goals different from Congress, the formula as implemented may be different from what Congress intended.

Relative to variant (a), variant (b) allows Congress to reflect its goals in the specification of the formula, giving it more control over the allocations. Under variant (b), Congress must be able to assess the performance of alternative formulas. Still, if there are a variety of ways of defining and estimat-

---

[1]Examples of the different arrangements are given in Chapter 1, in the section on "Alternative Approaches to Fund Allocation."

ing the components of the formula (e.g., need, capacity, effort), the implementation of the formula may be different from that envisioned by Congress.

Relative to variant (b), variant (c) allows Congress to be sure that implementation will be closer to what it envisioned. However, variant (c) requires more technical capability and understanding by Congress to ensure satisfactory performance. Under variant (c), Congress still delegates to the program agency the choice of data sources for estimating the inputs.

With variant (d), Congress retains full control over the allocation formula, the definitions of inputs, and the statistical data series used to calculate the allocations. It needs some expertise in statistical matters to evaluate and select candidate statistical data series.

Variant (e) does not require the use of statistical data in the allocation process; Congress specifies the amounts or shares for each recipient. This variant does not allow the allocations to reflect changes in social conditions over time, except through new legislation.

Recommendation 1. For each formula allocation program, an effective trade-off between congressional control and locus of expertise and monitoring must be found. As they have already done for many formula allocation programs, legislators should consider giving some flexibility to program agencies, especially in determining what data sources and procedures should be used to produce estimates of the components of allocation formulas.

PERIODIC EVALUATIONS

In the initial design stage and in subsequent revisions, formula designers are faced with a large array of choices, including:

- Which of the basic components—need, fiscal capacity, and effort—should be included in the formula?
- What conceptual variables should be used to represent those components?
- What data sources and procedures should be used to estimate those variables?
- How should the separate components be combined in the basic formula?
- What limits, if any, should be applied to the values of individual formula components or to formula outputs (recipient amounts or shares)?

Formula allocation programs are designed either to operate for an indefinite period of time (with annual allocation of funds appropriated for the program) or to operate for a specified number of years (usually four to six), at which time reauthorization of the program is considered. Formula designers and evaluators need to be aware of this longitudinal nature. Inevitably, there will be changes in the total funds available for the program, the distribution of need among aid recipients, and the nature and quality of the data sources available for use in estimating formula components. Trade-offs between stability of funding and adjustments to meet shifting needs should be considered. Modification of the formula or other program features may be indicated. Changes can be effected either by amending the initial authorization legislation or as part of the annual appropriations process. Changes effected by the latter can be made only for one year at a time; permanent changes require amending the authorizing legislation.

Recommendation 2. At reauthorization and possibly at other times, policy makers should evaluate whether a formula allocation program is performing as intended. Evaluations should include a study of how the relations between formula outputs (allocations) and inputs (measures of need, fiscal capacity, and effort) are affected by special provisions, such as hold harmless and small-state minimums. Evaluations should attempt to identify misallocations due to statistical variation or inherent bias in formula inputs and the degree of improvement in the accuracy of allocations that would be achieved by using improved inputs.

For some programs, a more comprehensive evaluation may be warranted to consider the quality of services delivered to program beneficiaries, the impact of those services, and program efficiency. Such assessments could be used to determine the net value to society of improving the accuracy of formula inputs or of revising the formula. For example, comparing the potential benefits of conducting a new survey or expanding a current survey to obtain more accurate estimates of need to the costs of data collection would show whether the overall improvement in performance of the formula allocation program justifies the investment in the survey or whether justification must depend on benefits outside the program. The ability to conduct such assessments successfully will depend to a considerable extent on how explicitly the goals of the formula allocation program have been defined in its authorizing legislation.

CHOICE OF VARIABLES, ESTIMATION PROCEDURES, AND DATA SOURCES

Many aid formulas include components that are intended to represent variation in need among recipient jurisdictions. However, some variables used as proxies for need are not closely linked to the resources needed to provide particular services, and developing good proxies is quite difficult. While there is general agreement that aid formulas should take account of differences in need, efforts to do this are often inadequate. Furthermore, statistical data series used to calculate the formula allocations can be updated and in many cases are revised as additional data become available or refinements are identified.

Recommendation 3. The formula design process should include evaluation of the trade-offs in timeliness, quality, costs, and other factors that are inherent in the selection of variables, the methods used to estimate them, and the data sources used to provide inputs. Specific considerations include:

— Whether to use direct estimates of a program's target population or need or to use proxies that may be more current, reliable, or available at less cost. For example, current estimates of total population by state (or population in a specific age range) might be an adequate proxy for some more difficult-to-measure population group.

— Whether to use estimates from a single source or model-based estimates that combine data from several sources. For example, the small-area estimates of income and poverty used in the Title I education program combine information from the decennial census with more timely data sources, such as the Current Population Survey and the federal income tax and Food Stamp programs.

— Whether to use the latest available data to compute each year's allocations, while recognizing that updating some series but not others may prove counterproductive.

— Whether, when modifications in data series or statistical estimation procedures are expected to produce a substantial change in allocations, to smooth the transition from the old to the new allocations.

ANTICIPATING BEHAVIORAL RESPONSES OF
AID RECIPIENTS

In the absence of legislated controls, there are many program-allowed strategies that recipient jurisdictions can adopt to maximize their benefits in a formula allocation program, and sometimes these strategies run counter to program goals. The most obvious example is substitution, whereby funds provided for a specific program are used to replace existing expenditures for that purpose, leaving total expenditures on the program at or near their original level. Other examples of unintended behavior include:

• When the measure of need depends on data reported by the recipient jurisdictions and reporting requirements are not clearly defined, some recipients may adopt definitions that work in their favor.

• When the formula includes current expenditure in the program area as an input, the amount shown in the books could be manipulated to maximize the jurisdiction's benefit without changing the amount actually spent.

• When the aid formula includes a representative tax system-style measure of fiscal capacity and a single jurisdiction accounts for almost all of a particular source of revenue (e.g., some varieties of oil in Saskatchewan in the Canadian Equalization Program), there can be a disincentive for the jurisdiction to tax that source, since any increase in the tax rate may be directly offset by a loss of equalization funds.

• When one jurisdiction is dominant in the state or the nation in which an aid program operates (e.g., the city of New York in the formulas for New York state aid to education), there may be incentives for that jurisdiction to influence overall state or national averages to its advantage.

Recommendation 4. Designers of formula allocation programs should evaluate the potential for unintended behavioral responses by recipient jurisdictions. Particular attention should be given to the possibility of substitution, to the ability of recipient jurisdictions to influence input data, and to the effect on calculated amounts that might result from particular actions by dominant jurisdictions. When formulas use data collected and provided by the recipient units, Congress and the relevant agencies should require use of standard definitions and should establish procedures to monitor and control quality.

SPECIAL FEATURES: HOLD-HARMLESS PROVISIONS

Many formula allocation programs have special features that modify the initial allocations determined by application of the basic formula. Hold-harmless provisions limit the extent to which recipients' aid amounts can decline from year to year, either in absolute terms or as a share of the total amount appropriated. Their effects on allocations depend on the amount of change permitted (a 100 percent hold harmless on shares guarantees no change in shares; a 100 percent hold harmless on amounts guarantees no reduction in amounts) and on the annual changes in total appropriations available for allocation (increase, decrease, level).

Hold-harmless provisions can limit disruptions in program administration and service delivery at state and local levels. However, they can also delay response to changing patterns of need. In the extreme case, a 100 percent hold-harmless provision coupled with no increase in program funding results in no changes from the previous year's allocations, regardless of changes in the indicators of need, fiscal capacity, and effort. In the presence of sampling variability in the estimation of formula components, hold-harmless provisions can cause a ratcheting-up effect that tends to favor smaller jurisdictions with less stable estimates. Moving averages, used alone or in conjunction with hold-harmless provisions, can reduce this ratcheting effect, but they also may delay responses to significant changes in the distribution of need.

Recommendation 5. The probable effects of hold-harmless provisions on allocations should be evaluated before including them in a formula, and their effects should be part of subsequent review. If undesired effects are identified, policy makers should consider such changes as loosening the hold-harmless constraint or introducing the use of moving averages—as an alternative means for reducing the volatility of allocations—to estimate formula components.

SPECIAL FEATURES: THRESHOLDS

Some allocation formulas include thresholds to ensure that certain designated funds are allocated only to those jurisdictions in which need is relatively great. For example, the concentration grant formula for the Title I education program includes a threshold on the number of poor children and the child poverty rate. With a threshold, a small change in estimated need, whether caused by statistical variation or a change in true need, can substantially affect the funding received by a jurisdiction. Another aspect of thresholds worth noting is that they may, given their all-or-nothing nature, raise the incentive for manipulation by potential fund recipients.

> **Recommendation 6.** The effects of a threshold on allocations should be evaluated both before it is included in a formula and subsequently whenever the formula is being evaluated. Evaluation should focus on how errors in estimated need cause jurisdictions to gain or lose funding erroneously and how fluctuations in estimated need affect the stability of allocations over time. If undesired effects of a threshold are discovered, policy makers should consider revising the formula to allow for a more gradual transition from no funding to above-threshold funding.

SPECIAL FEATURES: SMALL-STATE MINIMUMS

Many formula allocation programs have provisions for small-state minimums, which ensure that no state will receive less than a specified dollar amount or a specified share of the total allocation. A wide range of values have been used for small-state minimums. For example, the State Children's Health Insurance Program requires that each state receive at least $2 million annually, and the Substance Abuse Prevention and Treatment Block Grants program requires that no state receive less than 0.375 percent of the total amount allocated to the states.

> **Recommendation 7.** Formula designers should examine the degree to which proposed minimum amounts or shares produce departure from allocations based on estimated needs.

SIMULATIONS OF FORMULA ALLOCATIONS

When Congress develops or reauthorizes a formula allocation program, it is customary to run one or more currently available data series through alternative versions of the allocation procedure. Legislators and staff examine such "simulations" to see whether the results are satisfactory or whether the formula or data require modification. Sometimes program agencies, the General Accounting Office, or the Congressional Research Service provides assistance. For example, Brown (2002) and Riddle (2000) have presented results of simulations designed to show the effects of selected features of the allocation procedure for the Title I education program. To assist program recipients in their fiscal and budgetary planning, some agencies and organizations have published the results of simulations based on advance estimates of the formula inputs or on alternative formula features being considered by Congress.

The panel endorses simulation as an important tool to assess cross-sectional and longitudinal effects of alternative formulas, data inputs, and special features and to explore more general issues, such as the effects of alternative methods of combining formula components representing need, fiscal capacity, and effort. Simulations can shed light on the trade-offs faced in designing formulas, such as the trade-off between stability and responsiveness to need. Simulations have usually been cross-sectional, dealing with inputs and outputs for a single year; sometimes two consecutive years are considered, for example, to study the effects of a hold-harmless provision.

> **Recommendation 8.** Use of simulations when programs are being developed and when they are being reviewed prior to reauthorization should be expanded. Particular attention should be given to analyses of the effects of special features, such as hold-harmless provisions, caps, thresholds, and limits. Expanded analyses should include longitudinal studies to explore how statistical error in the input data series affects allocation patterns over time. Changes in funding levels, need distributions, and input data sources also indicate a need for evaluations of allocation processes, using simulation techniques, that are longitudinal rather than just cross-sectional. Simulations should be used to explore cross-cutting issues, such as choices among alternative measures of fiscal capacity and the relative merits of using hold-harmless provisions or moving averages to dampen the effects of large changes in the formula inputs.

INFORMATION ABOUT FEDERAL FORMULA ALLOCATION PROGRAMS

Formula allocation programs and procedures are often complex, and it is often necessary to consult several sources to achieve a full understanding of the history and current status of a particular program. The Catalog of Federal Domestic Assistance, maintained by the General Services Administration and made available at http://www.cfda.gov, provides an excellent starting point for learning about the features of specific formula allocation programs.

Recommendation 9. The General Services Administration should improve the utility of the Catalog of Federal Domestic Assistance by including in each program description:

— The name and current contact information of one or more individuals who have in-depth knowledge of the program, including, when relevant, the formula allocation procedures used.
— The functional areas (e.g., agriculture, education, health, and transportation) and subcategories that apply to the program. Enhancing users' ability to search the database flexibly (e.g., by multiple fields) would also be valuable.

INTERAGENCY COLLABORATION

There have been some beneficial informal contacts among staff with similar responsibilities in different program agencies. Such networking is desirable, and a more formal mechanism is needed. The Office of Management and Budget's Committee on Data Access and Confidentiality (http://www.fcsm.gov/cdac/index.html) provides an excellent model. Since its organization in 1996, the committee has made significant contributions to the development and widespread adoption of improved procedures for providing access to aggregate statistics and microdata, while preserving the confidentiality of individual information obtained for statistical purposes.

Recommendation 10. The Statistical Policy Office of the Office of Management and Budget should establish a standing Interagency Committee on Formula Allocations with the mission of disseminating general information about formula features and data sources to formula designers and program administrators and to conduct or sponsor research on relevant technical issues. Primary activities of the committee should include:

— Development of improved simulation procedures for use in the design and evaluation of allocation formulas and processes.

—Development of quality-control procedures to ensure that allocation procedures are carried out correctly.

—In cooperation with the General Accounting Office and the Congressional Research Service, development of a handbook on fund allocation formulas and processes. Such a handbook would provide an introduction to underlying concepts and practical considerations in the use of formulas to allocate funds. It would be valuable to people in the legislative and executive branches with direct or indirect involvement in the design and operation of formula allocation programs and could be used in training programs for various audiences. The panel offers a draft table of contents (see Appendix D).

In addition to the recommended primary activities, the committee might undertake or sponsor research on topics such as:

• The statistical properties of alternative methods for combining components in an allocation formula.

• Statistical issues that arise in using formula components designed to compensate for interstate differences in fiscal capacity, focusing primarily on the kinds of state data needed to estimate such components, the existing sources of such data, and the costs and benefits of developing new data sources. This review would compare per capita income, which is usually used for this purpose, and the Treasury Department's measure of total taxable resources. Other approaches, such as Canada's representative tax system, could be considered.

• Statistical issues that arise in using formula components designed to compensate for geographical variability in the cost of services to be provided by a program.

References

Advisory Commission on Intergovernmental Relations
  1977   Federal grants: Their effects on state-local expenditures, employment levels, wage rates. *The Intergovernmental Grants System: An Assessment and Proposed Policies.* Vol. A-61. Washington, DC: Advisory Commission on Intergovernmental Relations.

Bradbury, K.L., and H.F. Ladd
  1985   Changes in the revenue-raising capacity of U.S. cities, 1970-1982. *New England Economic Review* (March/April):20-37.

Break, G.F.
  1980   *Financing Government in a Federal System.* Studies in Government Finance. Second Series. Washington, DC: The Brookings Institution.

Brown, P.S.
  2002   Impact of Title I formula factors on school year 2000-01 state allocations. *Journal of Official Statistics* Special Issue (September):441-464.

Burnam, M.A., P. Reuter, J.L. Adams, A.R. Palmer, K.E. Model, J.E. Rolph, J.Z. Heilbrunn, G.N. Marshall, D. McCaffrey, S.L. Wenzel, and R.C. Kessler
  1997   *Review and Evaluation of the Substance Abuse and Mental Health Services Block Grant Allotment Formula.* RAND Drug Policy Research Center, MR-533-HHS. Santa Monica, CA: RAND.

Czajka, J.L., and T.B. Jabine
  2002   Using survey data to allocate federal funds for the State Children's Health Insurance Program (SCHIP). *Journal of Official Statistics* Special Issue (September):409-428.

Derbyshire, M.
  2001   Discussion on the paper by Smith, Rice, and Carr-Hill. *Journal of the Royal Statistical Society* 164(2):241.

Downes, T.A., and T.F. Pogue

1994 Accounting for fiscal capacity and need in the design of school aid formulas. In *Fiscal Equalization for State and Local Government Finance*, J.E. Anderson, ed. New York: Praeger Publishers.

2002 How best to hand out money: Issues in the design and structure of intergovernmental aid formulas. *Journal of Official Statistics* Special Issue (September):329-352.

Duncombe, W., and A. Lukemeyer

2002 Estimating the Cost of Educational Adequacy: A Comparison of Approaches. Unpublished paper, Maxwell School, Syracuse University.

Fairley, W.B., A.J. Izenman, and P. Bagchi

1990 Inference for welfare quality control programs. *Journal of the American Statistical Association* 85(411):874-890.

Fastrup, J.C.

2002 Measuring state performance in equalizing access to educational resources: The case of Rhode Island (1992-1996). *Journal of Education Finance*, forthcoming.

Gramlich, E.M., and H. Galper

1973 State and local fiscal behavior and federal grant policy. In *Brookings Papers on Economic Activity*, A.M. Okun and G.L. Perry, eds. Washington, DC: The Brookings Institution.

Hansen, M.H., and B.J. Tepping

1990 Regression estimates in federal welfare quality control programs. *Journal of the American Statistical Association* 85(411):856-863.

Kadamus, J.A.

2002 Formula allocation for schools: Historical perspective and lessons from New York state. *Journal of Official Statistics* Special Issue (September):465-480.

Kramer, F.D.

1990 Statistics and policy in welfare quality control: A basis for understanding and assessing competing views. *Journal of the American Statistical Association* 85(411):850-855.

Ladd, H.F.

1994 Measuring disparities in the fiscal condition of local governments. In *Fiscal Equalization for State and Local Government Finance*, J.E. Andersen, ed. New York: Praeger Publishers.

Ladd, H.F., and J. Yinger

1994 The case for equalizing aid. *National Tax Journal* 47(1):221-224.

Lindsey, L., and R. Steinberg

1990 *Joint Crowdout: An Empirical Study of the Impact of Federal Grants on State Government Expenditures and Charitable Donations.* NBER Working Paper No. 3226. Cambridge, MA: National Bureau of Economic Research.

Melnick, D.

2001 Précis for the Legislative Process and the Use of Indicators in Formula Allocations. Unpublished work, Dan Melnick Research, Inc.

2002 The legislative process and the use of indicators in formula allocations. *Journal of Official Statistics* Special Issue (September):353-370.

National Research Council

1999a   *Making Money Matter: Financing America's Schools.* Committee on Education Finance. H.F. Ladd and J.S. Hansen, eds. Washington, DC: National Academy Press.

1999b   *Equity and Adequacy in Education Finance: Issues and Perspectives.* Committee on Education Finance. H.F Ladd, R. Chalk, and J.S. Hansen, eds. Washington, DC: National Academy Press.

2000   *Small-Area Income and Poverty Estimates: Priorities for 2000 and Beyond.* Panel on Estimates of Poverty for Small Geographic Areas. C.F. Citro and G. Kalton, eds. Committee on National Statistics, Division of Behavioral and Social Sciences and Education. Washington, DC: National Academy Press.

2001   *Choosing the Right Formula: Initial Report.* Panel on Formula Allocations. T. Louis, T. Jabine, and A. Schirm, eds. Committee on National Statistics, Division of Behavioral and Social Sciences and Education. Washington, DC: National Academy Press.

Puma, M.J., and D.C. Hoaglin

1990   Food stamp payment error rates: Can state-specific performance standards be developed? *Journal of the American Statistical Association* 85(411):891-899.

Riddle, W., and P. Osorio-O'Dea

2000   *Where the Money Goes in Department of Education K-12 Programs.* CRS Report for Congress. Washington, DC: Congressional Research Service.

Siegel, M.

2001   Community Development Block Grant — Formula Funding. Unpublished paper presented to the Panel on Formula Allocations, Washington, DC, April 2001.

Smith, P.C., N. Rice, and R. Carr-Hill

2001   Capitation funding in the public sector. *Journal of the Royal Statistical Society* 164(2):217-257.

Spencer, B.D.

1985   Optimal data quality. *Journal of the American Statistical Association* 80:564-573.

State Controller's Office, State of California

2000   *School Districts Annual Report (1997-98 Fiscal Year).* Sacramento, CA: State Controller's Office.

2001   *Counties Annual Report (1998-99 Fiscal Year).* Sacramento, CA: State Controller's Office.

2002   Title I in California Programs and Policies web page. [Online]. Available at: http://www.cde.ca.gov/iasa/titleone/programs.html.

Substance Abuse and Mental Health Services Administration

2001   Substance Abuse Prevention and Treatment Block Grant. [Online]. Available at: http://www.samhsa.gov/programs/content/brief2001/01csaptblock.htm [October 11, 2002].

Suzara, F.B.

2002   A study on the formulation of an assessments scale methodology: The United Nations experience in allocating budget expenditures among member states. *Journal of Official Statistics* Special Issue (September):481.

Tannenwald, R.

    1999   Fiscal disparity among the states revisited. *New England Economic Review* (July/August):3-25.

    2001   Interstate Fiscal Disparity and Comparisons of Business Tax Climate for FY 1997. Unpublished paper, Federal Reserve Bank of Boston.

Taylor, M., S. Keenan, and J. Carbonneau

    2002   The Canadian Equalization Program. *Journal of Official Statistics* Special Issue (September):393-408.

U.S. Bureau of the Census

    1975   *Historical Statistics of the United States, Colonial Times to 1970, Part 2.* Washington, DC: U.S. Department of Commerce.

U.S. General Accounting Office

    1983   *Changing Medicaid Formula Can Improve Distribution of Funds to States.* GAO/GGD-83-27. Washington, DC: U.S. Government Printing Office.

    1990   *Federal Formula Programs: Outdated Population Data Used to Allocate Most Funds.* GAO/HRD-90-145. Washington, DC: U.S. Government Printing Office.

    1996   *Federal Grants: Design Improvements Could Help Federal Resources Go Further.* GAO/AIMD-97-7. Washington, DC: U.S. Government Printing Office.

    1999   *Formula Grants: Effects of the Adjusted Population Counts on Federal Funding to States.* GAO/HEHS-99-69. Washington, DC: U.S. Government Printing Office.

Weingroff, R.F.

    2001   For the common good: The 85th anniversary of a historic partnership. *Public Roads* 64(5):30-45.

Zaslavsky, A.M., and A.L. Schirm

    2002   Interactions between survey estimates and federal funding formulas. *Journal of Official Statistics* Special Issue (September):371-392.

Appendix A

Background Papers

These papers, which were prepared for the panel, are to be published in a Special Issue of the *Journal of Official Statistics.*

How Best to Hand Out Money: Issues in the Design and Structure of Intergovernmental Aid Formulas
Thomas Downes, Tufts University
Thomas Pogue, University of Iowa

The Legislative Process and the Use of Indicators in Formula Allocations
Dan Melnick, Dan Melnick Research

Interactions Between Survey Estimates and Federal Funding Formulas
Alan M. Zaslavsky, Harvard University
Allen L. Schirm, Mathematica

The Canadian Equalization Program
Michelle Taylor, Finance Canada
Sean Keenan, Finance Canada
Jean-Francois Carbonneau, Statistics Canada

Using Survey Data to Allocate Federal Funds for State Children's Health
Insurance Program (SCHIP)
> *John L. Czajka, Mathematica*
> *Thomas B. Jabine, Statistical Consultant*

WIC Funding Formula Evolution
> *Dawn Aldridge, U.S. Department of Agriculture*

Impact of Title I Formula Factors on School Year 2000-01 State Allocations
> *Paul Sanders Brown, U.S. Department of Education*

Formula Allocation for Schools: Historical Perspective and Lessons from
New York State
> *James A. Kadamus, New York State Education Department*

A Study on the Formulation of an Assessments Scale Methodology: The
United Nations Experience in Allocating Budget Expenditures Among
Member States
> *Felizardo B. Suzara, United Nations Statistics Division*

To provide background for its deliberations, the panel commissioned a
set of papers that appear as articles in a special issue of the *Journal of Official
Statistics.* Each article benefited from reviews by the guest editors, a subset
of panel members, and at least one outside referee. This special issue is a
joint activity of the Panel on Formula Allocations and the *Journal of Official
Statistics.*

The first three articles lay out how the formula allocation process
works; examine the underlying goals, roles, and structure of fund allocation
formulas; and describe the legislative development process and how formula
features, underlying data, and estimation procedures interact in producing
formula outputs. These articles are followed by U.S.-based and international case studies that serve to illustrate many of the issues raised in the
first three articles. The case studies are drawn from U.S. programs that
address children's health, women's and children's nutrition, and education;
from a Canadian program designed to reduce discrepancies in the fiscal
capacity of the provinces; and from the United Nations' dues assessment
procedures.

In the opening article, Thomas Downes and Thomas Pogue examine the design and structure of intergovernmental aid formulas. The authors discuss a wide range of issues that relate to alternative, often contradictory, aid objectives. In addressing how best to hand out money, they show how program goals can be optimally translated into aid formulas. Optimality relates to such goals as reducing fiscal disparities and making tax and expenditure activities of recipient governments geographically neutral; generating a more equitable distribution of tax burdens; or reducing inefficient provisions of a public service attributable to interjurisdictional spillovers. The authors assess the extent to which, in practice, formulas deviate from the ideal and examine the economic and social effects of these deviations.

In the second article, Dan Melnick focuses on how, in the context of broadly stated program goals, the legislative process influences the design of formula-based funds allocation. He looks at how the formula-based approach itself influences both the legislative process and government programs. The statistical challenges associated with formula allocation needs are complicated by operational realities that can affect the choices of statistical indicators ultimately used for formulas. The paper provides new insights into the process whereby policy makers and statisticians must fashion formulas that pass the test for face validity while generating the necessary political support.

Alan Zaslavsky and Allen Schirm provide a compliment to the Downes and Pogue paper. They describe how formula characteristics, data input sources, and statistical estimation procedures interact to determine funding allocations. Emphasizing allocation of U.S. federal funding to state and local entities, their central theme is that, while some consequences of these interactions are straightforward to predict, others are not. As a result, fund allocations are not always consistent with program goals. The authors simulate a series of multiyear scenarios to illustrate combinations of formula properties, data sources, and estimation procedures that are likely to produce allocations that don't line up with original intentions. They give special attention to problems caused by the introduction of new surveys for producing formula inputs.

The fourth paper leads off a set of program-specific case studies. Michelle Taylor, Sean Keenan, and Jean-Francois Carbonneau provide an overview of the Canadian Equalization Program, a program designed to narrow fiscal disparities among the provinces through intergovernmental aid. The basis of the transfers is to enable the relatively less prosperous and more prosperous provinces to provide roughly comparable levels of services. The authors discuss the historical and administrative background of the program and emphasize the central role of formulas in meeting program goals.

John Czajka and Thomas Jabine extend the volume's coverage of issues treated by Zaslavsky and Schirm by evaluating the use of survey data to estimate inputs for allocation formulas. The authors provide a comprehensive overview of the allocation process for the State Children's Health Insurance Program and recommend improvements. They discuss statistical problems created when allocations must be based on survey estimates that have large sampling errors.

Dawn Aldridge examines the history and formula features of the Special Supplemental Nutrition Program for Women, Infants, and Children (WIC). Aldridge traces how program goals have been manifested in a sequence of formulas, starting from the initial phase after the program was introduced in the early 1970s (when the congressional mandate emphasized expanding the number of eligible people reached by the program) to the 1990s, when the program had stabilized and program goals shifted toward an equitable distribution of resources across states. Aldridge shows how the rule-making process for the WIC program attempted to reflect these objectives.

Title I, Part A, of the Elementary and Secondary Education Act establishes grants to states and local educational agencies (LEAs) or school districts with disproportionate numbers of poor school-age children. Paul Brown analyzes how the interaction of factors such as the introduction of new data sources, use of hold-harmless guarantees, and political compromise has affected Title I allocations. For the 2000-2001 school year, he assesses how formula features interact to affect, sometimes in a contradictory manner, state-level and per child allocations. The case study illustrates "the tension that exists between the conflicting needs to target funds and to ensure funding stability for those LEAs that stand to lose as a result of new data."

In a second article on education funding, James Kadamus summarizes the development of school aid in his state. He communicates valuable lessons for those charged with using formulas to distribute aid at any level of government. He examines the historical context and the stated objectives of education aid in New York from the practitioner's perspective, describing the evolution of New York school aid formulas as incremental but "punctuated by occasional reforms." Kadamus discusses the effects on this evolution created by the often competing goals of increasing educational opportunities and improving financial stability as well as by problems of targeting aid to student needs, ongoing funding formula design and data quality challenges, and the federal role in education.

In the final article, Felizardo Suzara provides an international perspective. Unlike the foregoing articles, his describes a formula that allocates a tax obligation rather than a benefit. He shows how the United Nations (UN) uses formulas to allocate among member states the contributions required to finance its operations. Suzara describes how the UN's Committee on Contributions prepares the scale of assessments and advises the General Assembly on all aspects of its methodology "with a view to making it simple and transparent, stable and, most importantly, fair and equitable." In this context, in which capacity to pay can be calculated in a variety of ways, the formula is manifestly the result of extended negotiations and compromises among the participating stakeholders.

Appendix B

A Review of Twelve Large Formula Allocation Programs

To further our understanding of the statistical aspects of formula allocation procedures, the panel gave special attention to some of the largest federal programs and some others with special features. Federal agency officials responsible for several of these programs were invited to make presentations at panel meetings. In addition, the staff prepared program descriptions, using a standard format, for 12 programs, including the 10 largest in fiscal year 1999 and two others with features of particular interest. Box B-1 shows, as an example, the description that was prepared for the special education program. Each program agency was given an opportunity to review the description of its program, and several changes were made as a result of these reviews.

This appendix, which presents findings from an analysis of features of these 12 programs, has two sections. The first section describes several features of each of the 12 programs, focusing particularly on aspects of their formulas and allocation processes that are unusual in some way or that may be less than optimum for producing equitable distributions and efficient outcomes, or for which the rationale is not obvious. The second section provides some general observations about features that were common to some or all of the programs. Table B-1 lists the 12 programs included in the analysis and shows their total obligations in FY 1999. Together, they account for FY 1999 obligations of $190.5 billion or 79.3 percent of total obligations for the 169 formula allocation programs listed in Category A of the Catalog of Federal Domestic Assistance in 2000. The

---

Box B-1
Formula Allocation Program Description

CFDA number and name: 84.027 Special Education Grants to States

Subprogram (if applicable):

Basis for allocation: [   ] Amounts or shares in legislation
[x] Formula in legislation
[   ] Formula in regulations

Source of funds: [x] General revenues
[   ] Special (describe)

Recipients: [x] States
[   ] Other (describe)

Matching provisions: [x] No
[   ] Yes (describe)

Type of formula: [   ] Closed
[x] Other (describe)

The basic formula is closed, but because of a variety of caps and hold-harmless provisions, an iterative process is required to determine the final allocations.

Formula:

The "permanent formula," which first took effect in FY 2000, with FY 1999 as the base year, is:

$$A_{it} = A_{i\,99} + (A_t - \Sigma_i A_{i99})\,(0.85\,P_{1\,i}\,/\,\Sigma_i P_{1\,i} + 0.15\,P_{2\,i}\,/\,\Sigma_i P_{2\,i})$$

Prior to FY 2000, allocations were based on state counts of children served in their programs.

Definition of left-hand element:

A_{it} = amount allocated to state i in fiscal year t.

continued

Box B-1 Continued

Definition of right-hand elements:

Symbol	Definition	Source	Remarks
1. A_{i99}	Amount allocated to i^{th} state in base year.		
2. A_t	Total amount available for allocation to states in fiscal year t.		
3. P_{1i}	Total population in age range mandated by the state's program.	Most recent data satisfactory to Sec. of Education	Age ranges vary by state, max. range is 3 to 21.
4. P_{2i}	No. of poor children in age range mandated by state's program.	Most recent data satisfactory to Sec. of Education	See item 3.

Analysis of formula elements:

Need:

P_{1i} and P_{2i} are indirect estimators of need.

Capacity:

None.

Effort:

A_{i99} is a measure of effort because it was based on state counts of the number of children served.

Other:

None.

continued

Other formula features:

Upper and lower limits: [ ] No [x] Yes (describe)

Upper limit. (1) No state may receive more than an amount equal to the number of its children receiving special education services multiplied by 40 percent of the average per-pupil expenditure in U.S. public elementary and secondary schools. So far, appropriations have not been sufficient for this limit to take effect. (2) No state may receive more than its allocation for the previous year increased by the percentage increase in the total amount appropriated plus 1.5 percent.

Lower limit. See Hold-harmless.

Thresholds:　[x] No　[ ] Yes (describe)

Hold-harmless:　[ ] No　[x] Yes (describe)

As long as there is an increase in funds compared to the preceding fiscal year, no state may receive less than its allocation for that year. Additional provisions ensure that states will receive some minimum proportion of the increase in the amount appropriated for the current fiscal year.

State minimum:　[ ] No　[x] Yes (describe)

See Hold-harmless.

Remarks:

If there is a decrease in funds compared to the preceding year, but the amount is greater than the base year, the amount each state will receive is given by:

$$A_{it} = A_{i99} + (A_t - A_{99}) \times (A_{i(t-1)} - A_{i99}) / \Sigma_i (A_{i(t-1)} - A_{i99})$$

If the amount is less than the base year, base year allocations are ratably reduced.

Source: 20 U.S.C. 1411.

TABLE B-1   Twelve Large Formula Allocation Programs

Catalog Number[a]	Program	Total Obligations FY 1999 (*$billions*)
93.778	Medical Assistance Program (Medicaid)	111.1
20.205	Federal-Aid Highway Program	26.2
93.558	Temporary Assistance for Needy Families (TANF)	18.8
84.010	Title I Education	7.7
10.555	National School Lunch Program (Food portion)	5.5
84.027	Special Education Grants to States	4.3
93.767	State Children's Health Insurance Program (SCHIP)	4.2
93.658	Foster Care, Title IV-E	4.0
14.218	Community Development Block Grants Entitlement Grants	3.0
10.557	WIC (food portion)	2.9
93.959	Substance Abuse Prevention and Treatment Block Grants	1.5
66.458	EPA State Capitalization Grants	1.3
Total		190.5

[a]Catalog of Federal Domestic Assistance classification.

first 10 programs are those with the largest obligations in FY 1999. In addition, the substance abuse block grants program is discussed for its unique approach to equalizing fiscal capacity between states, as is EPA's state capitalization grants program, one of the few with numerical values of shares specified in legislation.

FEATURES OF INTEREST BY PROGRAM

Medical Assistance Program (Medicaid)

• Medicaid is by far the largest federal formula allocation program, accounting for 45 percent of total obligations for such programs in FY 1999. The key element of the formula is the federal medical assistance percentage (FMAP), which determines the proportions of state expenditures that will be reimbursed by the federal government. FMAP is also used, either directly or indirectly, in several other formula allocation programs, including three covered by this report: Temporary Assistance for Needy Families, the State Children's Health Insurance Program, and foster care. Therefore it is important to pay close attention to the structure and role of FMAP.

• The value of FMAP for each state is [1.00 − 0.45 (State PCI/National PCI)^2], where PCI stands for per capita income. FMAP is subject to the restriction that it cannot be less than 0.50 or greater than 0.83. A possible rationale for this formula is that the federal government should pay 55 percent of state Medicaid expenditures to a state whose per capita income is equal to the national per capita income. For states with per capita income below the national average, the federal government would pay a higher proportion, and vice versa. The lower limit of 50 percent was retained from a predecessor program that provided a flat matching rate of 50 percent to all states (U.S. General Accounting Office, 1983).

• At present, no states have per capita income so low that they are affected by the 0.83 upper limit on FMAP. However, several states with high per capita incomes are receiving 50 percent matching funds, more than they would receive if there were no lower limit.

• The FMAP formula uses the squared ratio of state to national per capita income. Within the 0.50 to 0.83 range, squaring enhances the effect of state variations from the national per capita income. Any state whose per capita income exceeds the national average by 5.4 percent or more receives a 50 percent match, whereas, if the ratio were not squared, this would occur only when the difference was 11.1 percent.

• The use of variable matching rates presumably represents an attempt to use federal funds to equalize (at least partially) the fiscal capacities of states to pay for the program. The formula uses per capita income as a proxy measure of fiscal capacity. Another option would be to use the Treasury Department's indicator, total taxable resources, which is used for this purpose in the substance abuse and mental health block grant programs.

• In recent years, the values of FMAP for Alaska and the District of Columbia have been set by statute at levels higher than those dictated by the formula.

Federal-Aid Highway Program

• The federal-aid highway program consists of several distinct subprograms, the largest of which use legislated formulas to apportion funds to the states. An unusual feature of the program is that all of the funding comes from the highway trust fund, which is a financing mechanism established by law to account for receipts collected by the federal government from motor fuel taxes; sales taxes on heavy vehicles, trailers, and tires; and use taxes on heavy vehicles. The portion of the funds from these taxes that goes into the highway trust fund is specified by law; currently, small portions go to the U.S. Treasury for deficit reduction and for cleanup of leaking underground storage tanks. The highway trust fund is divided into two separate accounts: the highway account, which supports the federal aid highway program, and the mass transit account, which supports the federal transit formula grants program.

•   Most of the distinct subprograms that make up the federal aid highway program have similar formulas,[1] which take the general form:

$$A_i = A \sum_j w_j M_{ij} / M_j$$

where

> A = amount available for the subprogram
> A_i = amount apportioned to state i.
> M_{ij} is a measure of need (such as vehicle miles, lane miles, or population), for state i, with $M_j = \sum_i M_{ij}$
> w_j is the weight for the jth measure of need, with $\sum_j w_j = 1$.

•   There are several questions that could be asked about the apportionment formulas for the subprograms. For example, how are the weights (w_j) associated with the measures of need chosen, and how well does each of the measures of need reflect the needs that are intended to be met by the program? For several of the measures used, such as vehicle miles and lane miles, the data used in the apportionments are provided by the state agencies that receive the apportioned funds, so procedures have been established to ensure that states do not manipulate the figures in an effort to increase their shares.

---

[1]For one of the larger programs, the high-priority projects program, there is no apportionment formula. The amount available for allocation each fiscal year is based on a percentage of the total authorized funding level for all high-priority projects over the life of TEA-21, pursuant to 23 U.S.C. § 117(b). Section 1602 of the authorizing legislation, P.L. 105-178, lists 1,850 specific projects, along with the dollar amounts authorized for them under this subprogram. Section 1603, P.L. 105-178, excludes several of these projects from the minimum guarantee calculation.

- In most of the programs, the apportioned funds are used by the states for specific projects for which they are required to provide matching funds. In most instances the federal share is 80 percent. Unlike the Medicaid program, the highway program has no provisions that attempt to equalize fiscal capacity among states.

- The legislation that reauthorized the program for FY 1998 to 2003 contains a table of state percentages adding to 100 percent, with a provision that for FY 1998 each state's apportionments over a specified set of subprograms, as a percent of such apportionments to all states, should equal the value shown in the table. Each fiscal year thereafter, these percents are modified to ensure that each state will receive a minimum of at least 90.5 percent of its percentage share of contributions to the highway trust fund account, based on the latest data available at the time of apportionment. The shares of states falling below that minimum return are increased and the shares of the remaining states are proportionately decreased so that the shares continue to total 100 percent (P.L. 105-178, Section 1104 or 23 U.S.C. § 105).

- The legislation authorizes as much funding as necessary, designated as minimum guarantee funds, to achieve the modified target state percentages. A portion of each state's share of these funds is added to the amounts apportioned to it under five other subprograms, all of which use formulas to apportion funds to the states. The remainder is made available to the states under the surface transportation program. The net result is that all of the apportionments are formula-driven, but the overall share of these apportionments received by each state is the numeric value specified in the law, as modified by the minimum guarantee provision.

Temporary Assistance for Needy Families (TANF)

- The TANF program was created by 1996 legislation designed to change the nation's welfare system into one requiring work in exchange for time-limited assistance. It replaced the Aid to Families with Dependent Children (AFDC) and Job Opportunities and Basic Skills Training (JOBS) programs. The goals of TANF are to promote work, responsibility, and self-sufficiency and to strengthen families. States operate their own programs, subject to certain statutory requirements.

- TANF block grants to the states are based on their expenditures for AFDC benefits and administration, emergency assistance, and the JOBS program during the period just prior to passage of the 1996 welfare reform legislation. There are three different options for measuring these historical expenditures; we are not aware of the rationale for making these choices available. The AFDC program, which matched state expenditures at a rate determined by the FMAP formula, was by far the largest of the three programs replaced, thus the amounts of TANF block grants depended indirectly on the FMAP formula. Therefore, to the extent that the use of FMAP reduced differences in fiscal capacity between states, it can be said that the TANF block grants also did this.

• Under the heading of maintenance of effort, states are required to maintain their own spending at 80 percent of FY 1994 levels. For states that meet certain work requirements, the mandatory state effort is reduced to 75 percent.

• The program includes several performance-related incentive and penalty provisions, many of which require the states to maintain detailed data collection and reporting systems.

The 1996 legislation provided for appropriation of $16.8 billion each year for the federal block grants through FY 2002. The deadline for reauthorization is September 2002, with the likelihood that new allocation formulas will be considered.

Title I Education

• Throughout the United States and especially in the South, there was considerable resistance to compliance with the Supreme Court's 1954 decision in *Brown v. Board of Education*, requiring desegregation of public schools. In the mid-1960s, the Johnson administration and Congress developed a carrot and stick approach to encourage compliance, with civil rights legislation providing the stick and the provision of federal funds for education, under Title I of the Elementary and Secondary Education Act, as the carrot. Title I funds were available only to jurisdictions complying with desegregation requirements.

• For most formula allocation programs, federal funds are allocated to the state agencies, which are then responsible for further distribution to subordinate jurisdictions. In this program, however, since school year 1999-2000, funds have been distributed by the U.S. Department of Education directly to school districts.[2]

---

[2] There are some exceptions for school districts with less than 20,000 population.

• One element of the formula, the state per pupil expenditure, is restricted to a range extending from 32 to 48 percent of the national average. The role of this element in the formula is ambiguous. It could be regarded as a relative measure of either need or effort. In either case, however, the fact that a statewide average is used means that within-state variations are not taken into account.

• The program has eligibility thresholds, with the result that a difference of one in the estimate of the number of eligible children or total school-age children for a school district can determine whether or not the district receives any funds. For example, to be eligible for a concentration grant, the estimated number of eligible children for a school district must be at least 6,500, or it must exceed 15 percent of the total estimated school-age population.

• In recent years, current estimates of number of eligible children by school district have been developed and updated biennially by the Census Bureau, with a model-based estimation procedure using data from the decennial census, the Current Population Survey, and administrative sources. This system replaced the earlier use of decennial census-based estimates, which were updated only every 10 years. However, for the past four fiscal years, Congress has enacted a 100 percent hold-harmless provision, so that lacking any significant increase in annual appropriations, the revised estimates have had little effect in shifting funds to areas where needs have increased more rapidly.

• The nature of the small-state minimum allocation is such that some states receive considerably more per eligible child than they would receive in the absence of this provision.

• Although this is the fourth largest federal formula allocation program, it accounts for only about 2 percent of total expenditures for public schools. Another 5 percent comes from other federal programs. Much larger amounts of state funds are distributed to school districts. The distribution of New York state funds for public elementary and secondary education equals about twice the total amount distributed to all states under the federal Title I program.

National School Lunch Program

• In simplest terms, this program pays the states a specified amount per lunch served to schoolchildren under the program. The amount paid per lunch (the national average price) varies according to whether the lunch is a paid, reduced price, or free meal. These national average prices are updated annually, based on the food away from home series of the consumer price index. Special prices have been established for Alaska and Hawaii because of the high costs of food and labor in those states. Schools are also entitled to receive a fixed value of commodities from the U.S. Department of Agriculture for each meal served; in some cases, schools can elect to receive cash instead of the donated commodities.

• The amounts paid to the states depend on eligibility determinations by the schools and on their reports of the numbers of lunches served in each category. The program regulations include a system of reports and audits in an attempt to ensure accuracy.

• One goal of the program is to promote high nutritional standards for lunches served to schoolchildren. There is an elaborate set of regulations covering this aspect.

• The special prices for Alaska and Hawaii established by legislation call attention to the fact that the Bureau of Labor Statistics does not produce state-level consumer price indices, so that price variations between states cannot be routinely included in allocation formulas.

Special Education Grants to States

• When this program was initiated by the Education for All Handicapped Children Act in 1975, annual allocations to the states were based on their certified counts of individuals being served under their programs, that is, a measure of effort. The current allocation procedure, based on the 1997 amendments to the act, is quite different. Each state receives its base year (FY 1999) allocation plus a proportion of the additional funds available, as determined by the state's share of two weighted quantities: total number of children in the age ranges mandated by the state's program and number of children in poverty in these age ranges. The weights are 0.85 and 0.15, respectively. The age ranges mandated by the individual state programs vary, with the maximum range for allocation purposes being from 3 to 21 years.[3]

• The quantities used in the current formula to allocate *additional* funds have elements of both need and effort. They are indirect measures of

---

[3]For some state programs, coverage extends to 22 or 23 years of age, but only ages 3 through 21 are considered for allocation purposes.

need because they represent the total population for which the eligible children constitute a subset, and they are measures of effort because they depend on the varying age ranges which the states have chosen to include in their programs. Recent increases in total appropriations for the program have the effect of giving greater weight to the need components of the basic formula. However, a complex set of minimum and maximum limitations on changes from year to year and from the base year delays responses of the allocations to changes in need and effort. For example, no state may receive more than its allocation for the previous year increased by the percentage increase in the total amount appropriated plus 1.5 percent, and no state may receive less than its allocation for the previous year increased by the greater of the percentage increase in the total amount appropriated minus 1.5 percent or 90 percent of the percentage increase in the amount appropriated. Such provisions may also create incentives for states to refrain from increasing the mandated age ranges for coverage by their programs, or even to reduce them.

- No state may receive more than an amount equal to the number of its children receiving special education services multiplied by 40 percent of the average per-pupil expenditure in U.S. public elementary and secondary schools. So far, appropriations have not been sufficient for this limit to take effect.

- At first glance, it might seem that the Census Bureau's biennial estimates of total school-age children and school-age children in poverty (the program estimates that were developed primarily for use in the Title I education program) might be ideal for use in this allocation formula. However, the panel was told in a presentation at its April 2001 meeting, this was not feasible because the age ranges for special education programs vary by state, and the Census Bureau's program does not produce estimates by single year of age. Hence, a source of data that is less accurate and up to date is being used.

State Children's Health Insurance Program (SCHIP)

- The goal of the program is to provide health insurance coverage for low-income children who are not covered by Medicaid or other kinds of insurance. Allocations to states are based on measures of need, consisting of the average of total low-income children and uninsured low-income children in the state, plus a cost factor based on average wages in the health services industry. Initially, the target population component of the allocation formula included only the number of uninsured low-income children. The total number of low-income children was subsequently added to avoid penalizing the states that were most successful in providing insurance coverage under their programs.

- Estimates, by state, of the population components of need are three-year moving averages derived from the Current Population Survey. Because of concerns about the high variability of state estimates used as inputs to the allocation formula, Congress appropriated $10,000,000 annually, starting in FY 2000, to expand the Current Population Survey sample to improve reliability. Even with the resulting increases in sample size, it was clear that the estimates by state would have relatively high sampling variability, so hold-harmless provisions were introduced to provide greater stability in the annual allocations.

- The formula for the state cost component of need is $SCF_i = 0.15 + 0.85\,W_i\,/\,W$, where W_i is the state's average wages in the health services industry. The inclusion of the constant in the formula attenuates the effect of state variations in wages on the allocations.

- At least 90 percent of state allotments must be used to reimburse the states for a specified proportion of their program expenditures for eligible children. The proportion for each state is the "enhanced" FMAP, calculated as:

$$FMAP(E)_i = FMAP_i + 0.3\,(1 - FMAP_i) = 0.3 + 0.7\,FMAP_i$$

with a minimum value of 0.65 (because the minimum value of FMAP is 0.50) and a maximum of 0.85 (which occurs when $FMAP \geq 0.786$). For further discussion of FMAP, see the description of the Medical Assistance Program (Medicaid) above. The matching rates for SCHIP are higher than those used for Medicaid, even though the latter program serves a needier population.

Foster Care-Title IV-E

- The purpose of the program is to help states provide proper care for children who need placement outside their homes. Like Medicaid, it is an open-ended entitlement program. The proportion of operating expenses covered by the federal government is determined by the value of FMAP for each state. In addition, the federal government pays 50 percent of administrative expenditures and 75 percent of training expenditures. For further discussion of FMAP, see the description of the Medical Assistance Program (Medicaid) above.

- As was the case for Medicaid, in recent years, the values of FMAP for Alaska and the District of Columbia have been set by statute at levels higher than those dictated by the formula.

Community Development Block Grants, Entitlement Grants Program

- The stated objectives of the grants provided by this program are to develop viable urban communities by providing decent housing and a suitable living environment and by expanding economic opportunities, principally for people with low and moderate incomes. Of the total annual appropriation, 70 percent is allocated to metropolitan cities and urban counties (entitlement areas), and 30 percent to the remaining areas of the states. The nonentitlement areas are covered by a separate program.

- The allocation process uses two different formulas to determine the proportionate shares for each entitlement area. The larger value is adopted for each area; then all shares are ratably reduced so that their sum is one. Each of the formulas has the general structure:

$$S_i = \Sigma_j w_j M_{ij} / M_j$$

where

S_i is the proportionate share for the ith entitlement area.

M_{ij} is the value of the jth measure of need (such as crowded housing units or population), for the ith entitlement area, with $M_j = \Sigma_i M_{ij}$

w_j is the weight for the *j*th measure of need, with $\Sigma_j w_j = 1$

The suitability of some of the measures of need is open to question. For example, the number of housing units built before 1940 may indicate a need for rehabilitation of old housing in some central cities, but not necessarily in all metropolitan cities and urban counties.[4]

---

[4]Greenwich, Connecticut, has been mentioned as a city that gets more than it should because this measure of need is included in the formula.

- The latest available decennial census data are used for several of the measures of need—crowded housing, old housing, and poverty population—so there is a long lag time before these measures are updated.
- This is one of the few programs for which funds are allocated directly to units other than states. The definitions of the metropolitan cities and urban counties are revised by the Statistical Policy Office of the Office of Management and Budget after each decennial census. Close attention is given to this process by jurisdictions that are "on the bubble" for qualifying as entitlement areas.
- At the panel's April 2001 meeting, the representative of U.S. Department of Housing and Urban Development, which administers this program, provided a seven-page handout that provided an exceptionally clear explanation of this relatively complex allocation process and some examples of how the calculations are carried out. This document could serve as a model for other programs (Siegel, 2001).

Special Supplemental Nutrition Program for Women, Infants, and Children (WIC)[5]

- The stated mission of the WIC program is "to safeguard the health of low-income women, infants, and children up to age 5 who are at nutritional risk by providing nutritious foods to supplement diets, information on healthy eating, and referrals to health care." One component of the program is the free distribution of infant formula to eligible infants.
- WIC is one of the programs for which it has been left to the program agency, in this instance the Department of Agriculture's Food and Nutrition Service (FNS), to develop the detailed allocation procedures for the program in accordance with the general requirements of the authorizing legislation. Following public reviews of the proposed formulas and procedures, they are incorporated into departmental regulations.
- With the help of a contractor, FNS has developed procedures for producing annual model-based estimates of the number of children under age 5 who are eligible for the program, by state. These estimates are used to calculate "fair shares" of the total amount available for food benefit grants to state agencies. However, targeting of appropriated funds based on these

---

[5] This program covers the food costs under WIC. Administrative costs are covered by a separate formula grant program.

current estimates of need has taken a back seat to stability provisions. A 100 percent hold-harmless provision, which is called a "stability grant," ensures each state agency of receiving at least as much as it did for the previous year (unless there is a reduction in the total amount available). Prior to a 1999 rule change, any additional funds available were used first for inflation adjustments. Only after these adjustments were fully funded were any remaining funds allocated to states not receiving their fair shares. As a result of the 1999 rule change, of the balance (if any) available for food benefits grants after allowing for stability grants, 80 percent is used for grants to cover increases in food costs due to inflation, and only the remaining 20 percent is allocated to states that are not receiving their fair shares. A proposed regulation published by the Department of Agriculture in 1998 had provided that the balance would be divided 50-50 between inflation grants and fair-share allocations. However, a great majority of the comments received recommended that a larger proportion be reserved for the inflation grants, leading to the 80-20 split in the final regulation.

Substance Abuse Prevention and Treatment Block Grants

• This program provides a good example of the issues that must be faced in evaluating the costs and benefits of efforts to develop improved measures of need. Proxy measures of need used in the current allocation formula have focused on persons ages 18 to 24, especially in urban areas, presumably because they were considered to be at highest risk for substance abuse. However, an analysis of data on substance abuse (including both drugs and alcohol) from the National Household Survey of Drug Abuse, which is funded by the Substance Abuse and Mental Health Services Administration's 5 percent share of the appropriation, suggested that the needs of smaller, less urban states were greater than indicated by these proxy measures (Burnam et al., 1997). The survey has recently been expanded so that it can provide direct estimates of prevalence and treatment need for the larger states and model-based estimates for all states. Such estimates offer a possible alternative to the proxy measures of need now in use.

• A small-state minimum of 0.375 percent of the total appropriation was introduced as part of the annual allocation procedure in FY 1999. If total population were the only basis for allocation, Wyoming, the smallest state, would be entitled to only 0.176 percent of the total (1999 population estimates), and several other states would be under the 0.375 percent minimum.

• This is the only large formula allocation program that uses the Treasury Department's estimates of total taxable resources (TTR) as a measure of states' relative fiscal capacities. The equalizing effect of TTR is attenuated by the imposition of an upper limit of 0.40.

• The formula uses a complex cost of services index that includes components for wages, rents, and supplies. Its effect on the allocations is attenuated by the imposition of a lower limit of 0.90 and an upper limit of 1.10.

• At various times during the life of the program, hold-harmless provisions have interfered with efforts to improve equity among the states (National Research Council, 2001:25).

Environmental Protection Agency's (EPA) State Capitalization Grants

• This program provides funds to states for the construction of new wastewater treatment and pollution control systems. Needs for projects in several eligible categories are determined by periodic clean water needs surveys conducted by EPA. Initially funds allocated to the states were used to pay for specific projects, but starting with FY 1988, pursuant to 1987 amendments to the Clean Water Act, the funds have been used to establish clean water state revolving funds, which are loaned to localities for projects. States are required to use their own funds for 17 percent of the total amount of the revolving funds.

• For fiscal years prior to 1988, state allotments were based on a combination of population and specific needs for wastewater treatment and water pollution control, as identified by EPA in its clean water needs surveys. However, in the 1987 amendments, the formula was dropped and numerical values of state shares were specified in the Clean Water Act. The allocation based on shares specified in the legislation has not changed since, although as various territories were phased out of the program, their share was reallotted among the states.

• EPA also has a clean water state revolving fund program (CFDA No. 66.468) for which annual allocations are determined by a formula based on infrastructure needs of public water systems based on the most recent drinking water needs survey, as opposed to shares specified in legislation.

GENERAL OBSERVATIONS

Based on the foregoing review of 12 major fund allocation programs, we now make some general observations about various features of U.S. federal formula allocation programs. Specific programs are referred to by their short titles.

Alternative Approaches to Fund Allocation

For most programs, the allocation formula is specified in the legislation. In addition, for some programs in this category, such as SCHIP, the data sources for each formula element are specified in precise detail in the legislation. For others, such as Title I education, the program agency has been allowed substantial leeway in developing estimates of formula components; however in that instance, the legislation required that the estimation methodology be reviewed by a panel of the National Research Council. For population elements in the formula for the special education program, the legislation specifies the most recent data satisfactory to the secretary of education.

Two of the 12 programs reviewed, highways and EPA state capitalization grants, had numerical values of state shares (proportions) specified in legislation. In the EPA program, a formula based on a periodic survey of clean water needs was replaced by legislated shares in FY 1988, and the legislated shares continue to be the basis for allocation in that program. In the highway program, which also uses shares specified in legislation, the final allocations can differ from those shares because of the requirement that each state receive a minimum amount equal to at least 90.5 percent of its share of contributions to the highway trust fund. For programs, including Medicaid and foster care, that make direct use of the FMAP formula to determine the proportion of a state's program expenditures to be covered by the federal government, rates in excess of those that would be indicated by the formula have been established legislatively for Alaska and the District of Columbia.

At the other extreme, for the WIC program, Congress established the general objectives in legislation and left it to the program agency, the Department of Agriculture's Food and Nutrition Service, to develop the allocation formula, which is set out in regulations published by the agency.

Equalizing Fiscal Capacity Among States

One goal that is either stated or implicit for several formula allocation programs, whether general or highly focused, is to equalize the capacities of the recipient jurisdictions to provide services to their citizens. To meet this goal, allocations must vary inversely with the recipients' fiscal capacities. When equalization is a goal, per capita income is normally used as a measure of fiscal capacity, the only alternative in U.S. federal programs being the Treasury Department's measure of total taxable resources by state, which is used in the substance abuse block grants program.[6]

Of the 12 programs reviewed, 4 had equalization provisions; for 3 of these 4 programs, the provisions operate by providing for varying "match rates," that is, the proportions of states' expenditures that are reimbursed by the federal government. In the Medicaid and foster care programs, match rates are determined by the FMAP formula, varying from 83 percent for states with the lowest per capita income to 50 percent for those with the highest per capita income.[7] As noted earlier, several states with high per capita incomes are receiving 50 percent matching funds, more than they would receive if there were no lower limit. In the SCHIP program, an "enhanced FMAP" formula leads to match rates that vary from 65 to 85 percent. The same states that receive 50 percent matching funds under FMAP receive 65 percent under SCHIP's enhanced FMAP.

For the substance abuse block grant program, there are no matching provisions, but total taxable resources, used as a measure of fiscal capacity, is a component of the formula for allocating appropriated funds to the states. Measures of need and geographic cost differentials enter directly into the allocation formula, and the measure of total taxable resources operates inversely, so that states with lower taxable resources receive higher shares.

The TANF program could be said to provide for some movement toward smaller disparities in fiscal capacity between states in a less direct way. The TANF block grants were based primarily on states' historical expenditures under the AFDC program, which used the FMAP formula to determine federal match rates.

---

[6]Some states, for example, New York, use measures of wealth based on assessed values of real estate in formulas for allocating state funds to school districts.

[7]At this time, no states are at the 83 percent level.

Measures of Need

For some programs that allocate a fixed total appropriation, a direct estimate of the number of people in each jurisdiction who are eligible for program benefits or services is a major element of the allocation formula. Examples include WIC (children under age 5 with family incomes less than 185 percent of poverty) and Title I education (children ages 5 to 17 in families with incomes below the poverty line). Both programs rely on model-based estimates of eligibles, using the latest data available from several sources. For WIC, estimates by state are updated annually, and for Title I education, estimates by school district are updated every other year.[8]

In other programs, estimates of population for various categories are used, along with other elements, as *indirect* indicators of need. The community development block grant program uses total population as an element in one of its two alternative formulas and population in poverty in both of them. Population also appears indirectly in two other indicators of need: overcrowded housing units and population growth lag. The substance abuse block grants program, in the absence of reliable state estimates of the number of substance abusers, uses weighted estimates of population for groups believed to be at highest risk. The special education program uses the mean of two population estimates: the number of persons in the age range served by each state's program and the number of poor children in those age ranges. The federal-aid highway program, in its national highway system subprogram, uses lane miles on principal arterial highways divided by population as one of four indicators of need.

The SCHIP program faces a difficult problem in trying to measure need. Initially, allocations were based on estimates of the low-income (below 200 percent of the poverty level) uninsured children by state. However, continuing this method of allocation would have penalized states that had been more successful in enrolling children in their health insurance programs, so the basis for the allocation was changed to the mean of estimates, by state, of the total number of low-income children and the number of low-income uninsured children. Further adjustments may be necessary in order to reflect future changes in the distribution of eligible and enrolled children by state.

---

[8]Note, however, that in both of these programs, efforts to improve targeting to current needs by obtaining the best estimates feasible of eligible populations have been undermined by the inclusion of 100 percent hold-harmless provisions in the allocation procedures (see discussion below).

Some programs use indicators of need other than population. One of the elements used in the community development block grant allocation formulas is an estimate of the number of housing units built before 1940. The highway program uses a variety of need indicators, including, for example, vehicle miles traveled on principal arterial routes (excluding the interstate system) and diesel fuel used on highways, in allocation formulas for its subprograms. Prior to the introduction of legislated shares in FY 1988, the allocation formula for the EPA state capitalization grants program was based in part on estimates of states' needs for wastewater treatment and water pollution control facilities, as identified in periodic surveys conducted by the EPA.

Geographic Cost Differentials

If the needs of states and other jurisdictions are to be established in dollar terms, it seems reasonable that geographic cost differentials should be taken into account in establishing the level of need in each area. In programs like Medicaid, in which varying proportions of state program expenditures are reimbursed by the federal government, such differentials are, at least to some degree, taken into account automatically. Of the programs that allocate fixed total appropriations each year, two include cost factors in their formulas. The SCHIP allocation formula includes a cost factor based on mean annual wages in the health services industry by state. The formula for the substance abuse block grants program has a more complex cost factor, with elements representing the costs of rents, services, and supplies. In each of these programs, there are restrictions that prevent the cost factor from having its full potential effect on the allocations. The formula for the SCHIP cost factor is:

$$SCF_i = 0.15 + 0.85\, W_i / W$$

where W_i and W are average wages in the health sector for the state and the country, respectively. The constant factor attenuates the effect of this element on the allocation. In the substance abuse block grants program, the cost of services index for a state cannot be less than 0.9 or more than 1.1.

In the school lunch program, presumably in recognition of higher food costs in Alaska and Hawaii, the Department of Agriculture has set higher average lunch prices for those two states. One might ask whether data on food costs for all states should be used to set prices that vary by state. One obstacle to doing this is that the Bureau of Labor Statistics does not routinely provide data on retail and wholesale prices by state.

Ostensibly because of the higher cost of health services, the value of FMAP for Alaska has been legislatively increased over what it would be if the standard formula were used. The value of FMAP for the District of Columbia has also been increased by legislation.

Combining Multiple Elements in Formulas

It will be evident by now that many programs use allocation formulas that combine several different elements of need, fiscal capacity, and effort. As noted earlier, the allocation formulas for subprograms of the highway program determine each state's share of the amount available for the subprogram by taking a weighted average of the state's share of each of several elements of need:

$$A_i / A = \Sigma_j \, w_j \, M_{ij} / M_j$$

where
  A = amount available for the subprogram
  A_i = amount allocated to state i
  M_{ij} is a measure of need (such as vehicle miles, lane miles, or population) for state i, with $M_j = \Sigma_i M_{ij}$
  w_j is the weight for the jth measure of need, with $\Sigma_j w_j = 1$

A similar approach, which could be called "weighted average of shares," is used to combine elements of need in the alternative allocation formulas for the community development block grant program. However, other programs, including special education and SCHIP, use a "share of weighted averages" approach to accomplish the same purpose:

$$A_i / A = \Sigma_j \, w_j \, M_{ij} / \Sigma_i \Sigma_j \, w_j \, M_{ij}$$

The substance abuse block grants program uses a complex two-stage process for combining elements. In the first stage, elements of need are combined using the weighted average of shares approach. Separate elements of cost are combined in similar fashion. In the second stage of the process, indices representing population (need), cost, and fiscal capacity for each state are multiplied and the products summed over all states to determine the shares, i.e.:

$$A_i / A = N_i\, C_i\, F_i / \Sigma_i\, N_i\, C_i\, F_i$$

This brief look at methods of combining multiple elements in allocation formulas suggests two questions that deserve further study:

1. What is an appropriate basis for determining the relative weights that should be given to different elements of need?
2. What are the relative merits of the three alternative methods of combining elements that we have described: the multiplicative approach, the weighted average of shares, and the share of weighted averages?

Stability Versus Targeting Current Need

A matter of great concern, for both legislators and recipients of formula grant funds, is how the allocations to each jurisdiction change from year to year, especially in absolute terms. Legislators for areas whose amounts or shares decline are likely to face hard questions from their constituents. Unexpected or unpredictable declines in federal program funding can cause difficulties for state and local program administrators, for example, school officials planning their budgets for the coming school year. Still, most fund allocation programs are designed to meet fairly specific needs and to equalize, at least in part, the fiscal capacity of states to meet those needs. As needs and fiscal capacities change, one would expect allocations to be responsive to those changes. Except for open-ended programs like Medicaid and foster care, there is clearly a trade-off between stability and improved targeting of current needs, especially if annual program funding remains level or declines.

One approach to better targeting of funds is to improve the formula inputs, primarily by updating the data used to estimate needs and fiscal capacity. Sometimes improved estimators, such as the model-based estimates that have been developed for the Title I education and WIC programs, can be developed and introduced into the formula allocation process. However, these estimates, although on the average they may reflect needs more accurately than the inputs previously used, are still subject to error, which can be relatively large in some instances, such as the Title I education program, which requires estimates by school district, and the

SCHIP program, which requires state estimates for a narrowly defined subset of the total population.

In order to maintain some degree of stability, the allocation procedures for several programs have (or have had at various times in the past) *hold-harmless provisions* that guarantee that each recipient will receive, at a minimum, a specified proportion of the prior year's amount.[9] The specified proportion may be 100 percent or it may be less or more than 100 percent. In some programs, hold-harmless provisions remain the same from year to year; in others they have been established for a limited period, especially at times when revised formulas were being introduced. In some programs, if there has been an increase in appropriations from the prior year, only the increase has been used to bring recipients closer to their fair shares based on updated estimates.

In 2 of the 12 programs reviewed, hold-harmless provisions have clearly undermined efforts to improve the targeting of allocations to current needs. In the Title I education program, a model-based estimation procedure, with estimates updated biennially, has replaced the earlier use of decennial census-based estimates, which were updated only every 10 years. However, for the past four fiscal years, Congress has enacted a 100 percent hold-harmless provision, so that lacking any significant increase in annual appropriations, the revised estimates have had little effect in shifting funds to areas where needs have increased most rapidly. In the WIC program, in which similar steps have been taken to improve estimates of need, there is a 100 percent hold-harmless provision, called a "stability grant." Of the balance (if any) available for food grants after allowing for stability grants, 80 percent is used for grants to cover increases in food costs due to inflation, and only the remaining 20 percent is allocated to states that are not receiving their fair shares as determined from current estimates of need.

Prior to the introduction of a 100 percent hold-harmless provision in the Title I education program, the legislation covering basic grants included a partial hold-harmless provision based on a step function, with higher rates for jurisdictions with higher poverty. Areas with 30 percent or more poor school-age children were guaranteed at least 95 percent of the prior year's grant; areas with 15-30 percent poor school-age children were guaranteed 90 percent; and those with fewer than 15 percent poor school-age

---

[9]In some instances, the total amount available may be insufficient to meet the hold-harmless guarantee; in such cases allocations to all participants are "ratably reduced" to add to the total available funds.

children were guaranteed 85 percent. A difference of one person in the estimate of poor school-age children or total school-age children for a jurisdiction could therefore make a substantial difference in the size of the grant for that jurisdiction.

The special education program allocation rules ensure that as long as there is an increase in funds compared with the preceding fiscal year, no state may receive less than its allocation for that year. Additional provisions ensure that states will receive some minimum proportion of any increase in the amount appropriated for the current fiscal year. The SCHIP program has a hold-harmless provision that applies to shares rather than amounts. Recent legislation provided that, starting with FY 2000, no state's share could be less than 90 percent of its share for the preceding fiscal year, or less than 70 percent of its FY 1999 share.

Upper and Lower Limits

Several formula allocation programs place upper and lower limits on one or more formula components, so that these components are allowed to vary only within restricted ranges, with the result that some jurisdictions receive either more or less than they would have received if these limits did not apply. A notable example of these kinds of limits is the restriction of FMAP to a range between 50 and 83 percent. As noted earlier, no states are currently affected by the upper limit, but several of the states with the highest per capita incomes benefit substantially from the 50 percent lower limit, both in the Medicaid program and in other programs that rely on FMAP (or its enhanced version) to determine matching percentages.

Other examples of limits, by program, are:

- *Federal highway program.* No state can receive less than 90.5 percent of its estimated contributions to the highway trust fund.
- *Title I education.* State per pupil expenditure, which is multiplied by the estimated number of eligible children to provide an estimate of need, is restricted to a range between 80 and 120 percent of the national average per pupil expenditure.
- *Special education.* No state may receive more than an amount equal to the number of its children receiving special education services multiplied by 40 percent of the average per-pupil expenditure in U.S. public elementary and secondary schools. So far, appropriations have not been sufficient for this limit to take effect. In addition, no state may receive more than its

allocation for the previous year increased by the percentage increase in the total amount appropriated plus 1.5 percent.

- *SCHIP.* From FY 2000 on, no state's share can exceed 145 percent of its FY 1999 share.
- *Substance abuse block grants.* The cost of services index for a state cannot be less than 0.9 or more than 1.1, and the fiscal capacity index for a state cannot exceed 0.4.

The rationales for some of these limits appear fairly obvious. The 40 percent limitation for special education is based on a long-range target, set for the program when it was first enacted, that the federal government should eventually pay that proportion of the costs of special education. The federal highway program distributes funds collected from user taxes, and it seems reasonable that each state should receive at least some minimum proportion of the taxes attributable to it.

The reasons for other limit provisions, including those placed on the values of FMAP, are less obvious. The general impression conveyed is that formula developers did their best to develop equitable and effective allocation formulas and processes, taking into account needs, fiscal capacities, and levels of effort, but may have then been led, either because of concerns about data quality or by political considerations, to depart from their initial formulations.

Small-State Minimums

Of the 12 programs reviewed, 5 have provisions that guarantee a minimum amount or share to every state. The SCHIP program guarantees a minimum of $2 million, about 0.047 percent of the total appropriation for FY 2000, to every state. The highway and EPA state capitalization grants programs each guarantee a minimum share of 0.5 percent to every state, and the substance abuse block grants program guarantees a minimum of 0.375 percent. The specification of the state minimum for basic grants in the Title I education program is not quite so simple. Each state must receive a minimum of the smaller of (a) 0.25 percent of the total grants to states, or (b) the average of (1) 0.25 percent of total state grants, and (2) 150 percent of the national average grant per eligible child, multiplied by the estimated number of eligible children in the state. What this works out to is that if the state's proportion of U.S. eligible children is equal to or greater than .00167, it will receive a share of at least 0.25 percent. If the proportion is less than .00167, the state's share will be somewhat smaller.[10]

Although none of these programs uses total population as an indicator of need, it may still be of interest to observe that according to the 2000 census there are seven states that had less than 0.25 percent of the total U.S. population, with Wyoming, the smallest, having 0.16 percent. There are seven additional states that have more than 0.25 but less than 0.50 percent of the total. At least in some of the programs, provisions for small-state minimums ensure that grant amounts per person exceed the national average. One can see two kinds of justifications for these provisions, one practical and one political. Most programs require states to incur expenses to set up programs to administer the receipt and use of federal grant funds, and some of the costs may be more or less fixed regardless of a state's population. The other, political consideration is that small states have disproportionate representation in the Senate and their votes are needed to pass authorization and appropriation legislation for the programs.

Target Allocation Units

Of the 12 federal funds allocation programs reviewed, 10 allocate funds to states, often with special provisions for allocations to territories and American Indian tribes. The exceptions are the Title I education and community development block grants programs. For the Title I education program, which initially allocated funds to state education agencies and later to counties, the U.S. Department of Education now allocates funds directly to school districts. Clearly, it is much more difficult to develop precise estimates of the number of eligible children, which is the key component of the allocation formula, for school districts rather than for counties or states. But this is not a new requirement; previously it was up to the states to allocate the funds they received to school districts, and they used a variety of estimation procedures to do this. Shifting the burden to the federal government appears to have reflected the view of Congress that the federal program agency, the U.S. Department of Education, could do a better job.[11]

---

[10]For a detailed analysis of the effects of the small-state minimum, in conjunction with other formula features, on the Title I education allocations, see Brown (2002).

[11]As noted earlier, Congress may have had some reservations about this change, because states were given the option to make the allocations themselves for all school districts with populations under 20,000.

The community development block grants program allocates funds to metropolitan cities (central cities of metropolitan areas and other large cities) and large urban counties, collectively known as entitlement areas, and to states for the remaining nonentitlement areas. As noted earlier, the definitions of the metropolitan cities and urban counties are revised by the Statistical Policy Office of the Office of Management and Budget after each decennial census. Close attention is given to this process by jurisdictions that are "on the bubble" for qualifying as entitlement areas. The nature of these allocation units is such that data for several of the measures of need included in the two alternative allocation formulas, including poverty population, number of overcrowded housing units, and number of housing units built prior to 1940, are only available from the most recent decennial census. It might be appropriate in this kind of situation to consider the trade-offs involved in basing the allocations on measures of need that may not be quite as directly connected to the program goals, but for which current data could be more readily obtained.

Data Inputs Provided by Recipients

Much of the data for states and other areas that are used in allocation formulas comes from the Bureau of the Census and other federal statistical agencies, for example, the Bureau of Economic Analysis and the Bureau of Labor Statistics, or program agencies, for example, the Internal Revenue Service. However, for several fund allocation programs, data for important formula elements are compiled and provided by the same state agencies that operate the programs that the federal grants are intended to support. In these programs, precautions are needed to ensure that the data are compiled by the states according to clearly defined definitions and procedures, and that audits and other quality control methods are used to monitor adherence to those definitions and procedures.

For the Medicaid and foster care programs, the allocations (federal matching funds) are based on reports by the states of their eligible expenditures in state operated programs. Given the size of the Medicaid program especially, regulations (probably quite complex) for reporting and quality-control procedures to monitor accuracy undoubtedly exist, yet there have been reports of gaming by the states to increase their payments. The special education program initially based allocations on reports by states on the number of students served by their programs, but for FY 2000 this approach was replaced by a formula that depends on federal estimates of total children and children in poverty in the age ranges served by their programs. This change in the allocation process may have been due in part to congressional concerns that some states may have been enrolling some students in special education programs when it was not appropriate.

Several of the subprograms in the highway program use formulas with elements, such as lane miles and vehicle miles, for which data are provided by the states. In the Title I education program, the data on per pupil expenditures are reported by the states to the U.S. Department of Education. At the April 2000 Workshop on Formulas for Allocating Program Funds, it was reported that California had recently changed its method of counting school attendance (the denominator of per pupil expenditures) in an effort to increase its share. Unlike most states, it had previously defined attendance to include excused absences, thus putting itself at a disadvantage in relation to states that were not including them (National Research Council, 2001:22).

REFERENCES

Burnam, M.A., P. Reuter, J.L. Adams, A.R. Palmer, K.E. Model, J.E. Rolph, J.Z. Heilbrunn, G.N. Marshall, D. McCaffrey, S.L. Wenzel, and R.C. Kessler
   1997   *Review and Evaluation of the Substance Abuse and Mental Health Services Block Grant Allotment Formula.* RAND Drug Policy Research Center, MR-533-HHS. Santa Monica, CA: RAND.
National Research Council
   2001   *Choosing the Right Formula: Initial Report.* Panel on Formula Allocations. T. Louis, T. Jabine, and A. Schirm, eds. Committee on National Statistics, Division of Behavioral and Social Sciences and Education. Washington, DC: National Academy Press.

Appendix C

Sources of Information

In the course of its work, the Panel on Formula Allocations used many sources of information to obtain an overview of formula allocation programs and to explore pertinent features of specific programs. The panel focused primarily on programs of the U.S. federal government, but some attention was given to state-funded programs, especially aid to education, to programs in other countries (primarily Canada and Australia), and to international programs. In the last category, the panel looked at the system for determining dues payments by members of the United Nations, a system with many of the same features observed in formula-based programs for allocating funds.

This appendix describes the sources of information that we were able to identify and use. We hope this summary will be useful to others who have a general interest in the subject or are seeking information about specific programs. Formula allocation programs, and the processes involved in creating them, can be exceedingly complex. Often there may be only a few people who have a full understanding of their history and the manner in which the formulas operate. Therefore, it may be necessary to go to several sources to get the whole story.

U.S. FEDERAL PROGRAMS

General-Purpose Sources

Catalog of Federal Domestic Assistance

The Catalog of Federal Domestic Assistance (CFDA) is a database of all federal programs available to various entities, including state and local governments, American Indian tribal governments, territories, organizations, and individuals. The catalog is compiled by the U.S. General Services Administration (GSA) and is updated biannually, in June and December. The entire catalog is available to the public at http://www.cfda.gov. The catalog was very useful to the panel as the basis for an overview of federal formula allocation programs and as a starting point for more intensive analysis of a subset of these programs.

The programs in the CFDA are categorized by 15 types of assistance, of which Category A, "formula grants," is the most pertinent to the panel's work. The catalog describes the programs in Category A as "allocations of money to States or their subdivisions in accordance with distribution formulas prescribed by law or administrative regulation, for activities of a continuing nature not confined to a specific project." The 2001 version lists 172 formula grant programs in Category A, funded by nearly every federal department and a wide range of agencies within the departments. Over the course of the panel's work, GSA has reorganized this category, adding some programs from other classifications and removing a few that it determined did not fit the definition of formula grant. Also, in late 2001, GSA added a 2001 Formula Report to the CFDA web site, listing 161 programs with "assistance formulas." That report includes several programs that are not listed in Category A. These were added to arrive at a count of 180 federal formula allocation programs for fiscal year 2000.

For each program, the CFDA provides a description in standard format. Items for formula grant programs that were of special interest to the panel included:

- Program number and name
- Federal agency
- Authorization (identifies relevant legislation)
- Objectives
- Uses and use restrictions

- Eligibility requirements
- Formula and matching requirements
- Obligations (estimated for current and next fiscal year; actual for previous fiscal year)
- Information contacts (including web site address)

Earlier in its work, the staff developed a database to assist the Panel on Formula Allocations in its review of the statistical aspects of fund allocation formulas and procedures using the FY 1999 version of the CFDA. Data elements included: federal agency responsible for the program, functional areas (one or more of 20 categories used in the catalog), total obligations for FY 1999, and several identifiers, including a 5-digit identification number and the name, postal and e-mail addresses, and telephone number of a contact person in the federal agency responsible for the program. This database was used to prepare summary descriptive statistics of the number of programs and total obligations, classified by agency and functional area, and to provide a listing of programs by agency and one of programs in order by size (obligations).

The amount of information about formulas and associated allocation procedures that is contained in the CFDA program descriptions is often quite limited, and anyone wanting more detailed information would need to use other sources, as discussed below. The catalog allows the user to approach the individual programs in a number of ways. Users can search by category, functional area, department, and agency, among others. This is useful but has its limits, since the site does not cross-reference. Most important, the individual program descriptions do not list the functional areas for the programs; users of the CFDA web site would have to search by functional area to categorize them in that way. In some instances when we contacted agencies to request more information about a program, we found that the contact information was not current. Initially, we found that a few of the programs listed in Category A were not really formula grant programs and, conversely, that a few formula grant programs had been classified in other categories.[1]

---

[1]The director of the division of GSA that is responsible for the CFDA was very helpful in identifying the misclassified programs so that we could more accurately identify the set of programs that was of interest to the panel.

Formula Report to the Congress

The *Formula Report to the Congress*, produced by GSA through 1999 in response to a congressional mandate, was a listing of all federal formula allocation programs, with more detailed information than that provided by the CFDA. However, in 1996 Congress conducted a review of all recurring reports it required from the executive branch. Most of the reports reviewed, including the *Formula Report*, were listed for termination effective December 1999, allowing a three-year period for members to review and revise the list. No objections to terminating the *Formula Report* were heard, and effective December 20, 1999, Public Law 98-169 was amended to delete the requirement for the report. Therefore, some information in the *Formula Report*, including obligations and contact information, may be dated and of limited use. The final report was published in 1999.

The *Formula Report* had five sections. The first, the agency program index, categorized each program by U.S. department and federal agency. The applicant eligibility index listed each program by department and noted who is eligible to apply. Applicants included individuals, local and state governments, U.S. territories, Indian tribal governments, and non-profit organizations. The report also grouped programs into functional categories and subcategories, similar to the CFDA functional areas. The program descriptions were detailed descriptions of each formula allocation program. The report concluded with an appendix listing the resource person for each program.

Each program description provided the reader with a formula narrative, and some also included the mathematical structure of the formula, in addition to definitions of the formula components. This level of detail is very useful to anyone interested in the mechanisms of the individual formulas. In some cases, the program descriptions in the *Formula Report* contained a great deal more information about the individual programs and their formulas than the CFDA. For instance, the formula description for the Cooperative Extension Service (CFDA No. 10.500) included the entire formula narrative that appears in the legislation and a detailed mathematical rendering of the formula. Descriptions of the National School Lunch Program (10.555), and the Community Development Block Grants/ Entitlement Grants (14.218) provided similarly detailed information about the formulas for those programs. However, not all of the programs' formulas were listed in such detail.

Federal Funds Information for States

Federal Funds Information for States (FFIS) is a nonprofit data clearinghouse that gives states a facility to analyze data and help develop positions on legislation. Funded by the National Governors Association and the National Conference of State Legislators, it tracks approximately 230 grant programs that account for 90 to 95 percent of all funding that states receive from the federal government.

The database defines programs in various ways, including function (agriculture, health, etc.); congressional committee; CFDA number; recipients (states, local government, other); discretionary or mandatory funds; and formula or competitive grant program. Since the primary mission of the clearinghouse is to track and report on the fiscal impact of federal budget and policy decisions on state budgets and programs, the database is not constructed around formula features and does not contain detailed information on program formulas. However, it will run alternative formula constructions for their sponsors by request.

In addition to maintaining the program database, FFIS regularly reports on important grant-in-aid program issues through their *Issue Briefs*. These briefs analyze the state-by-state impact of program, formula, and data changes on grant-in-aid funding. They also publish *The Billion Dollar Club Series*, which focuses on the 44 federal grant-in-aid programs in the database that receive an appropriation exceeding $1 billion. Although FFIS is not specifically designed to provide detailed information about formulas, it can be a useful tool for research on formula allocation programs.

Program-Specific Sources of Information

To further its understanding of the statistical aspects of formula allocation procedures, the panel focused attention on a few of the larger federal programs and some others with unusual features. Federal agency officials responsible for several of these programs were invited to make presentations at panel meetings. In addition, the panel staff prepared program descriptions, using a standard format, for 12 programs, including the 10 largest in fiscal year 1999 and two others that had features of particular interest.

Acquisition of detailed information about formulas and allocation procedures for these 12 programs required the use of several sources of information in addition to the general-purpose sources described in the preceding section. Among the sources consulted were:

- *Legislation.* The legislation authorizing a formula allocation program often contains the formula to be used. Relevant legislation can be identified in the "Authorization" category of the CFDA program description. Sometimes the legislation also specifies the data sources to be used for specific formula elements. Occasionally the annual appropriations legislation will include provisions that affect the allocation process, such as a new hold-harmless procedure.

- *Regulations.* For some programs the specific allocation formula or the data sources used as inputs to the formulas have been developed or determined by the program agency, following general program goals and objectives specified in the authorizing legislation. The detailed formulas and procedures are normally published as federal regulations.

- *Agency web sites.* The program agencies' web sites often provide extensive information about the programs, including the allocation formulas and procedures used, as well as links to other sources of information. Frequently they identify agency staff who can be contacted for more information. Web site addresses are included in the CFDA program descriptions.

- *Agency documentation.* Some agencies have prepared detailed materials about their formula programs. The Federal Highway Administration produces "Financing Federal-Aid Highways," which describes the basic process by which federal highway funds are allocated. The U.S. Department of Labor offers "Grants to States for Costs of Administration of Unemployment Insurance Laws," describing funding for administrative costs of state unemployment programs.

- *Research reports.* Several program agencies have sponsored research designed to evaluate and improve formula allocation procedures and the quality of input data. Some examples of research publications include a report from the Panel on Estimates of Poverty for Small Geographic Areas (National Research Council, 2000) for the Title I education program, the RAND Corporation report on the substance abuse and mental health block grant programs (Burnam et al., 1997), a Mathematica Policy Research report on the estimation of target populations for the WIC program (Schirm, 1995), and reports by the Urban Institute (Blumberg et al., 1993) and the JWK International Corporation (Ellett, 1978) on the Medicaid matching formula. In addition, there have been many informative independent evaluations of funding formulas and allocation procedures by the General Accounting Office, the Congressional Research Service, and other organizations.

STATE-FUNDED AID PROGRAMS

Many programs allocate state funds to cities, counties, and other governmental units. The panel focused its attention on two states: California, for which we examined a wide variety of state aid programs (see Chapter 7), and New York, for which we focused on state aid to education (Kadamus, 2002). State aid to education is important because it covers nearly half the total cost of public elementary and secondary education, whereas only about 10 percent is covered by federal funding. One useful general-purpose source of information on state aid to education is the National Center for Education Statistics compendium, "Public School Finance Programs of the United States and Canada, 1998-99," which is available only on a CD-ROM or via the Internet (http://nces.ed.gov/edfin/state_finance/StateFinancing.asp). Also informative were two reports produced by the National Research Council's Committee on Education Finance (National Research Council, 1999a, 1999b).

OTHER COUNTRIES AND INTERNATIONAL ORGANIZATIONS

In the United States, the General Revenue Sharing program of the 1970s and 1980s represented an approach to intergovernmental financing sharply different from that now in force, with its multiplicity of federal grant programs, each targeting specific policy goals. Under general revenue sharing, unrestricted federal funds were allocated to nearly 39,000 local governmental units on the basis of a complex formula with elements representing need and fiscal capacity. Other countries, notably Canada, Australia, and the United Kingdom, have unrestricted revenue-sharing programs of long standing, and panel members considered it useful to review the main features of those programs. A paper on Canada's equalization program, with its formula-based allocation of federal funds to seven provinces, was commissioned and presented at the panel's fifth meeting (Taylor et al., 2002). Australia's program of income tax sharing with its states was of particular interest because of the role played by the Commonwealth Grants Commission, a permanent body established to provide advice to the Commonwealth about the principles and allocation procedures for these grants (for a detailed description of the Commission and its history, see Commonwealth Grants Commission, 1995). Useful references on recent trends in the United Kingdom's formula allocation programs include Smith et al. (2001) and the discussion that follows it.

The United Nations and other international organizations are also in the business of allocating resources to their constituents, national governments. The panel commissioned a paper, also presented at its fifth meeting, describing the evolution and current status of the UN's scale of assessments (Suzara, 2002). As noted earlier, the assessments procedure turned out to have many of the same features observed in formula-based programs for allocating funds.

REFERENCES

Blumberg, L., J. Holahan, and M. Moon
  1993 *Options for Reforming the Medicaid Matching Formula.* Washington, DC: The Urban Institute.
Brown, P.S.
  2002 Impact of Title I formula factors on school year 2000-01 state allocations. *Journal of Official Statistics* Special Issue (September):441-464.
Burnam, M.A., P. Reuter, J.L. Adams, A.R. Palmer, K.E. Model, J.E. Rolph, J.Z. Heilbrunn, G.N. Marshall, D. McCaffrey, S.L. Wenzel, and R.C. Kessler
  1997 *Review and Evaluation of the Substance Abuse and Mental Health Services Block Grant Allotment Formula.* RAND Drug Policy Research Center, MR-533-HHS. Santa Monica, CA: RAND.
Commonwealth Grants Commission
  1995 *Equality in Diversity: History of the Commonwealth Grants Commission,* 2nd ed. Canberra: Australian Government Publishing Service.
Ellet, C.
  1978 *Analysis of the Federal Medical Assistance Percentage (FMAP) Formulas, Volume I: Executive Summary.* Annandale, VA: JWK International Corporation.
Kadamus, J.A.
  2002 Formula allocation for schools: Historical perspective and lessons from New York State. *Journal of Official Statistics* Special Issue (September):465-480.
National Research Council
  1999a *Making Money Matter: Financing America's Schools.* Committee on Education Finance. H.F. Ladd and J.S. Hansen, eds. Washington, DC: National Academy Press.
  1999b *Equity and Adequacy in Education Finance: Issues and Perspectives.* Committee on Education Finance. H.F Ladd, R. Chalk, and J.S. Hansen, eds. Washington, DC: National Academy Press.

2000 *Small-Area Income and Poverty Estimates: Priorities for 2000 and Beyond.* Panel on Estimates of Poverty for Small Geographic Areas. C.F. Citro and G. Kalton, eds. Committee on National Statistics, Division of Behavioral and Social Sciences and Education. Washington, DC: National Academy Press.

Schirm, A.L.
1995 *State Estimates of Infants and Children Income Eligible for the WIC Program in 1992.* Washington, DC: PriceWaterhouseCoopers LLP.

Smith, P.C., N. Rice, and R. Carr-Hill
2001 Capitation funding in the public sector. *Journal of the Royal Statistical Society* 164(2):217-257.

Suzara, F.B.
2002 A study on the formulation of an assessments scale methodology: The United Nations experience in allocating budget expenditures among member states. *Journal of Official Statistics* Special Issue (September):481.

Taylor, M., S. Keenan, and J. Carbonneau
2002 The Canadian Equalization Program. *Journal of Official Statistics* Special Issue (September):393-408.

Appendix D

Handbook on Fund Allocation Formulas and Processes

On several occasions the panel discussed its proposed activities with representatives from federal and state agencies, the legislative branch, and nongovernmental organizations. Development of a handbook that would provide an introduction to underlying concepts and practical considerations in the use of statistical formulas to allocate funds elicited considerable enthusiasm. The panel developed a table of contents for such a handbook; however, we did not have the resources to produce it.

As stated in our recommendations chapter, the panel sees that there is a need for such a handbook, which should be produced under the auspices of the Office of Management and Budget's Office of Statistical Policy. The primary audience would be federal and state legislators and their staffs, but it should also be valuable to agency officials responsible for administration of fund allocation programs and to nongovernmental organizations with interests in these programs. It could also be an attractive component of graduate-level courses in public administration and finance.

PROPOSED TABLE OF CONTENTS

1. Introduction
    1.1   Purposes and scope
    1.2   Intended audience
    1.3   Uses
        1.3.1   Developing a new formula
        1.3.2   Analyzing/revising an existing formula
        1.3.3   Periodic evaluations

2. An overview of fund allocation programs
    2.1   History and current status
        2.1.1   An early example: The Morrill Act
        2.1.2   General revenue sharing
        2.1.3   A statistical summary of current federal programs
        2.1.4   Formula allocations of state funds
        2.1.5   International perspective
    2.2   The parties involved
        2.2.1   Congress
        2.2.2   Program agencies
        2.2.3   First-level recipients
        2.2.4   Individual beneficiaries
        2.2.5   Advocacy groups
    2.3   Alternative approaches to fund allocation
        2.3.1   Amounts specified in legislation
        2.3.2   Specific formula in legislation
        2.3.3   Goals in legislation; formula developed by program agency
    2.4   Types of formula allocations
        2.4.1   Single-pass, mathematical expressions
        2.4.2   Iterative procedures
        2.4.3   Matching and cost-sharing provisions

3. Program goals
    3.1   Desired outcomes
        3.1.1   Close the gap between need and effort
        3.1.2   Treat equals equally
        3.1.3   Encourage spending on targeted services
        3.1.4   Fair treatment of communities
    3.2   Target population
    3.3   Services provided

4. Basic features of formula allocation programs
    4.1   Target allocation units
           4.1.1   Multilevel allocations
    4.2   Ultimate beneficiaries
    4.3   Frequency and timing of disbursements
    4.4   Provisions for administrative costs
    4.5   Program rules

5. Components of allocation formulas
    5.1   Measures of need
           5.1.1   Workload
           5.1.2   Geographic cost differentials
    5.2   Measures of fiscal capacity
    5.3   Measures of effort
    5.4   Interactions among components

6. Special features of formula allocations
    6.1   Thresholds and other eligibility criteria
    6.2   Limits
    6.3   Hold-harmless provisions and caps
           6.3.1   Applied to total appropriation
           6.3.2   Applied only to increase in appropriation
           6.3.3   Moving averages as an alternative
    6.4   Step functions
    6.5   Bonuses and penalties
    6.6   Interaction of special features with size of and changes in
           program appropriations

7. Data sources for estimating formula components
    7.1   Decennial censuses
    7.2   Household surveys
    7.3   Other statistical programs
    7.4   Administrative records
           7.4.1   Aggregate data
           7.4.2   Individual records, e.g., student record systems
    7.5   Factors to consider in choosing data sources
           7.5.1   Conceptual fit
           7.5.2   Level of geographic detail available
           7.5.3   Timeliness

Appendix E

Participants in Panel Workshops
and Meetings

Dawn Kimberly Aldridge, Office of Analysis, Nutrition, and Evaluation
Charles Alexander, U.S. Census Bureau
Susan Binder, Federal Highway Administration
Paul S. Brown, U.S. Department of Education
John Czajka, Mathematica Policy Research, Inc.
Thomas Fanning, New York State Health Department
Jerry Fastrup, U.S. General Accounting Office
Gregory Frane, U.S. Department of Education
Marcia Howard, Federal Funds Information for States
James Kadamus, New York State Department of Education
Daniel Kasprzyk, National Center for Education Statistics
Sean Keenan, Finance Canada, Equalization and Policy Development
Jerry Keffer, U.S. Census Bureau
Cindy Long, Food and Nutrition Service
David McMillen, U.S. House of Representatives
Daniel Melnick, Consultant, Committee on National Statistics
Tim Ransdell, California Institute for Federal Policy Research
Wayne Riddle, Library of Congress
Marjorie Siegel, U.S. Department of Housing and Urban Development
William Sonnenberg, National Center for Education Statistics
Felizardo B. Suzara, United Nations Statistics Division
Cynthia Taeuber, University of Maryland, Baltimore County
Karen Wheeless, U.S. Census Bureau
Albert Woodward, Substance Abuse and Mental Health Services Administration

Appendix F

Biographical Sketches of
Panel Members and Staff

THOMAS A. LOUIS (*Chair*) is professor of biostatistics, Johns Hopkins Bloomberg School of Public Health. His statistical research focuses on Bayesian methods; applied areas include biomedicine, the environment, and public policy. He is coordinating editor of *The Journal of the American Statistical Association*, a member of the Committee on National Statistics (CNSTAT), on the board of the Institute of Medicine's (IOM) Medical Follow-up Agency, and on the executive committee of the National Institute of Statistical Sciences. He was on the IOM Panel to Assess the Health Consequences of Service in the Persian Gulf War and was on the CNSTAT Panel on Estimates of Poverty for Small Geographic Areas. He is a fellow of the American Statistical Association and of the American Association for the Advancement of Science, an elected member of the International Statistical Institute. From 1987 to 1999 he headed the Department of Biostatistics at the University of Minnesota. He has a Ph.D. in mathematical statistics from Columbia University.

MARIA ALEJANDRO (*Project Assistant*) is a staff member of the Committee on National Statistics. She is currently working on projects on the design of nonmarket accounts, cost of living index, and confidentiality and data access. Previously, she worked at University of California, San Francisco, and the U.S. Department of Housing and Urban Development Bridge project.

GORDON BRACKSTONE is assistant chief statistician responsible for statistical methodology, computing, and classification systems at Statistics Canada. From 1982 to 1985 he was the director-general of the methodology branch at Statistics Canada, and previously he was responsible for surveys and data acquisition in the Central Statistical Office of British Columbia. His professional work has been in survey methodology, particularly coverage assessment and estimation in censuses, and in the management of data quality in statistical agencies. He is a fellow of the American Statistical Association and an elected member of the International Statistical Institute. He has B.Sc. and M.Sc. degrees in statistics from the London School of Economics.

VIRGINIA A. de WOLF (*Study Director*) was, during the course of this study, a senior program officer on the staff of the Committee on National Statistics. In the early 1990s she served as the study director of the panel that authored *Private Lives and Public Policies: Confidentiality and Accessibility of Government Statistics*. Her areas of research interest are confidentiality and data access as well as statistical policy. Previously, she has worked at the U.S. Office of Management and Budget, the Bureau of Labor Statistics, the National Highway Traffic Safety Administration, the U.S. General Accounting Office, and the University of Washington (Seattle). She has a B.A. in mathematics from the College of New Rochelle and a Ph.D. from the University of Washington (Seattle) in educational psychology with emphases in statistics, measurement, and research design.

THOMAS A. DOWNES is associate professor of economics at Tufts University. His research focuses in part on the evaluation and construction of state and local policies to improve the delivery of publicly provided goods and to reduce inequities in the delivery of these services, with particular attention paid to public education. He has also pursued research that considers the roles of the public and private sectors in the provision of education. He has advised policy makers in several states; an example of this advisory work is his contribution to *Educational Finance to Support High Learning Standards,* the final report of a symposium sponsored by the New York State Board of Regents. He has a B.A. from Bowdoin College (1982) and a Ph.D. from Stanford University (1988).

LINDA GAGE is the liaison to demographic programs at the California Department of Finance. She represents California in federal and professional forums and evaluates the effect of various demographic and statistical programs on the state. Previously, she served as the California state demographer for two decades and in other positions within the Department of Finance since 1975. She is a member of the American Statistical Association, the Population Association of America, and the National State Data Center Program Steering Committee. She has served on the U.S. Secretary of Commerce's Decennial Advisory Committee since 1995. She has B.A. and M.A. degrees in sociology, with emphasis in demography, from the University of California, Davis.

MARISA A. GERSTEIN (*Research Assistant*) is a staff member of the Committee on National Statistics. She is currently working on projects on methods for measuring discrimination, nonmarket accounts, and elder abuse and neglect. Previously, she worked at Burch Munford Direct, a direct mail company, and the National Abortion and Reproductive Rights Action League. She has a B.A. in sociology from New College of Florida.

HERMANN HABERMANN is director of the United Nations Statistics Division. In this position, he has the responsibility of providing leadership and coordination in the development of international statistical standards, technical cooperation, methodological work and in data collection activities. Previously, he held positions as chief statistician and deputy associate director for budget in the Office of Management and Budget. In addition to his statistical experience and knowledge, he has extensive knowledge of the federal statistical system. He has a B.S. from the Pratt Institute and a Ph.D. from the University of Wisconsin, Madison, in statistics.

THOMAS B. JABINE is a statistical consultant who specializes in the areas of sampling, survey research methods, statistical disclosure analysis, and statistical policy. Recent clients include the Committee on National Statistics, the National Center for Health Statistics, and several other statistical agencies and organizations. He was formerly statistical policy expert for the Energy Information Administration, chief mathematical statistician for the Social Security Administration, and chief of the Statistical Research Division of the U.S. Census Bureau. He has provided technical assistance in sampling and survey methods to several developing countries for the United Nations, the Organization of American States, and the U.S. Agency for International Development. His publications are primarily in the areas of sampling, survey methodology, and statistical policy. He has a B.S. in mathematics and an M.S. in economics and science from the Massachusetts Institute of Technology.

CHRISTOPHER MACKIE is a program officer with the Committee on National Statistics (CNSTAT). In addition to working with this panel, he is working on projects involving nonmarket economic accounting and data access and confidentiality. He was study director for the CNSTAT panel that produced *At What Price: Conceptualizing and Measuring Cost-of-Living and Price Indexes*. Prior to joining CNSTAT, he was a senior economist with SAG Corporation, where he conducted a variety of econometric studies in the areas of labor and personnel economics, primarily for federal agencies. He has held teaching positions at the University of North Carolina, North Carolina State University, and Tulane University. He is author of *Canonizing Economic Theory*. He has a Ph.D. in economics from the University of North Carolina.

ALLEN L. SCHIRM is a senior fellow at Mathematica Policy Research, Inc. Formerly, he was Andrew W. Mellon assistant research scientist and assistant professor at the University of Michigan. His principal research interests include small-area estimation, census methods, and sample and evaluation design, with application to studies of child well-being and welfare, food and nutrition, and education and training policy. He is currently an associate editor of *Evaluation Review*. He served on the Committee on National Statistics Panel on Estimates of Poverty for Small Geographic Areas and is currently a member of its Panel on Research on Future Census Methods. He is a member of the American Statistical Association's section on survey research methods working group on technical aspects of the Survey of Income and Program Participation. He has an A.B. in statistics from Princeton University and a Ph.D. in economics from the University of Pennsylvania.

BRUCE D. SPENCER is a professor of statistics and faculty fellow in the Institute for Policy Research at Northwestern University. His interests include the interactions between statistics and policy, demographic statistics, and sampling. He chaired the Statistics Department at Northwestern from 1988 to 1999 and 2000 to 2001. He directed the Methodology Research Center of the National Opinion Research Center (NORC) at the University of Chicago from 1985 to 1992. From 1992 to 1994 he was a senior research statistician at NORC. At the National Research Council he served as a panel member with the Mathematical Sciences Assessment Panel, and the Panel on Statistical Issues in AIDS Research; as a staff member he served as study director for the Panel on Small Area Estimates of Population and Income. He has a Ph.D. from Yale University.